SEA KING

SEA KING

Airlife
England

ACKNOWLEDGEMENTS

I would like to extend my sincere thanks to the many people who so generously contributed towards this book and to the Sea King aircrews who helped make the photographs possible. I would particularly like to thank the following:

Westland Helicopters Limited for the use of their photographs and in particular to Sue Eagles for all her support and assistance in the preparation of this book.

Mr Bob Turner for all his expert knowledge on the Sea King and for compiling all the Sea King data and serial numbers used in the book.

Lt Commander C. P. West, RN, Command Public Relations Officer, FONA and to all those on both 845 and 707 Squadrons at RNAS Yeovilton for all their help.

Lt Commander L. A. Port, MBE, RN, the Commanding Officer of 846 Naval Air Squadron for allowing me to join them in the Gulf and Kurdistan. To all the personnel of 846 Naval Air Squadron, past and present for all their kind assistance.

Commander T. J. Eltringham, OBE, RN, Commanding Officer, Commando Helicopter Operations and Support Cell for his support past and present.

Commander V. Sirett, OBE, RN (Retd), Community Relations Officer, RNAS Yeovilton.

Lt Commander Nigel North, RN, Commanding Officer of 848 NAS.

Lt Commander Mike Caws, RN, Editor of *Flight Deck*, the Journal of Naval Aviation, for access to his Falkland Conflict photographic library.

826 Naval Air Squadron and Lt Paul Gray, RN, 'C' Flight 826 NAS at RNAS Culdrose for all their help.

The Fleet Air Arm Museum at RNAS Yeovilton for all their assistance and for the use of their photographs.

Mr Michael J. Hill, Command Public Relations Officer, RAF Strike Command, and to all at RAF Mount Pleasant and 78 Squadron, Falkland Islands.

Squadron Leader Simon Turner, RAF, the Commanding Officer, RAF Sea King Training Unit and Fl Lt Steve Martin, RAF, for all their help.

Wing Commander Martin Mayer, RAF, ETPS and Squadron Leader John Taylor, RAF, at A&AEE Boscombe Down.

Lieutenant P. A. Cunnison, Public Relations Officer, RNAS Culdrose.

Lieutenant E. A. Waddell, Royal Australian Navy, Public Relations Officer, RAN Air Station NOWRA.

CONTENTS

INTRODUCTION

Originally designed by Sikorsky, the Sea King within eight months of its introduction in 1959 had captured all of the world's major helicopter speed records and then proceeded to secure its place in history as the most versatile, reliable and enduring maritime helicopter in the world. The civil variant of the Sea King, the S-61, in 1961 became the first commercial twin turbine helicopter to be certified for passenger operations and for instrument (IFR) flight by the Federal Aviation Agency and quickly became one of the world's most popular commercial helicopters.

Having obtained the Sea King production licence from Sikorsky in the early 1960s, Westland Helicopters Ltd began anglicising the basic design not only to fill the Royal Navy's ASW helicopter requirement but also adapted the aircraft for overseas customers to fulfil the search and rescue and troop transport roles. Their continued policy of updating and improving the Sea King has resulted in numerous Westland Sea King variants which include the latest Royal Navy Sea King HAS Mk6 and the export Westland Advanced Sea King.

In July 1991, the Sea King celebrated 21 Years Front Line Service with the Royal Navy. It was also the year that the entire front line strength of the Royal Navy's troop lifting Sea King helicopter force was deployed into the deserts of Saudi Arabia along with ship-based Sea Kings for 'Operation Granby'. Within a few weeks of the Gulf War Royal Navy Sea Kings were sent to Kurdistan for 'Operation Haven' and to Bangladesh for 'Operation Mana' where the Sea King was used in its humanitarian role. In the SAR role the Royal Air Force Sea King HAR Mk3 continues to provide long range Military and Civilian search and rescue throughout the UK and the Falkland Islands.

April 1992 also saw the tenth anniversary of the Falklands Conflict in which the Sea King proved so vital to the success of the operation, both at sea and on land. Both in the Falklands War and the deserts of Saudi Arabia during the Gulf War, and later in the mountains of Kurdistan, Sea King availability remained in excess of 90 per cent, a tribute to the helicopter's strength, reliability and to the maintainers who look after them.

In 1970 the Sea King helped to transform Anti-Submarine Warfare (ASW) and since that time has saved thousands of lives in its Search and Rescue (SAR) role as well as proving outstanding as a land-based troop transport helicopter.

In September 1991 it was announced that Westland and IBM had been selected to supply 44 EH-101 Merlin helicopters to the Royal Navy as their next generation ASW, ASUW and long-range SAR helicopter. The Sea King will continue in Royal Navy service for many years to come operating in the AEW, ASW, Commando and Search and Rescue roles deployed aboard aircraft carriers, Royal Fleet Auxiliary ships, Type 22 towed array frigates and from shore bases.

Patrick H. F. Allen
March 1993

SECTION 1
THE CONCEPT

By the 1960s Westland Helicopters had already forged a successful relationship with Sikorsky, building several of their helicopters under licence. These included the Sikorsky S-51 'Dragonfly', the Sikorsky S-55 'Whirlwind' and the S-58 'Wessex'. The anglicised S-58 now known as the Westland Wessex HAS Mk1 proved extremely successful. Fitted with a Napier Gazelle turboshaft engine, the Wessex provided the Royal Navy with their first day and night capable search and strike anti-submarine warfare (ASW) helicopter.

The success and versatility of the Wessex was soon noticed by the Royal Navy Commando Helicopter Squadrons who quickly replaced their Whirlwind Mk7s with the new Wessex Mk1. The Wessex Mk1 was basically similiar to the ASW version, except for the removal of the Flight Control System (FCS) and sonar equipment. The success of the Wessex in Fleet Air Arm service led Westland to continue their development programme and on 18 January 1962 the first Westland Wessex HC Mk2/HU Mk5 flew. The fitting of two Rolls-Royce Gnome turboshaft engines effectively doubled the helicopter's capabilities. The RAF ordered the Wessex HC Mk2 to fill their Tactical Helicopter needs and the Royal Navy, impressed with the advantages of additional power plus an emergency single engine capability, ordered the helicopter for their Commando Air Squadrons. The Royal Navy ASW Wessex HAS Mk1 was also the subject of continuous development by Westland and in 1967 the Westland Wessex HAS Mk3 entered service.

Although pleased with the Wessex HAS Mk1, the Fleet Air Arm wanted a long-range ASW helicopter that could work autonomously within the Fleet. Fitted with an 'Ecko' lightweight search radar, larger, more powerful Plessey 195 Sonar and a 'Duplex' fully Automatic Flight Control System (AFCS), the Wessex HAS Mk3 was the first ASW helicopter to be able to work independently of the Fleet both day and night. The new AFCS could take the helicopter from lift-off through a complete ASW sortie. With the AFCS married to the 'Ecko' search radar and a Marconi Doppler Navigation system which continually updated the helicopter's position, it was the first time that the Wessex could operate at night, without the assistance of a ship's radar service. Capable of combining the search and strike role, the Wessex HAS Mk3 had one drawback, that was the old problem of too much weight and a lack of power. Even fitted with an uprated Napier Gazelle 165 engine the extra weight of the radar plus new sonar equipment, coupled with lightweight homing torpedoes made sortie times much too short.

At an early stage both the United States Navy and Royal Navy knew they must have a larger twin-engined hunter-killer ASW helicopter capable of carrying greater payloads for longer periods. The speed in which the Soviet Union had been building and deploying both conventional and nuclear submarines into the world's oceans in the 1950/60s increased the urgency for such a helicopter. In the mid-1950s Sikorsky had designed a helicopter to fulfil a US Navy requirement. This was for a hunter-killer anti-submarine warfare helicopter which would have twin engines and be capable of long-range operations, day and night in all weathers. The US Navy awarded Sikorsky a contract to build their S-61 on 24 December 1957 and the first flight of the new S-61 took place on 11 March 1959. Once in service, the S-61 was redesignated the SH-3A Sea King. In 1966 the SH-3A was uprated to the SH-3D incorporating more powerful 1400 shp GE T58-GE-10 engines, structural strengthening, larger fuel tanks, uprated transmission and an increase in gross weight.

Soon after the first flight of the Sikorsky HSS2/S-61 in 1959, Westland concluded a licence agreement with Sikorsky to use the basic airframe and transmission to develop their own anti-submarine helicopter for the Royal Navy. In October 1966 the first of four Sikorsky SH-3Ds (G-ATYU-XV370) arrived in the UK and was flown from Bristol Docks to Yeovil by Westland Chief Test Pilot Slim Sears. Three more SH-3D airframes (XV371-XV372-XV373) arrived at Yeovil from Sikorsky and in less than three years the first production Sea King HAS Mk1 (XV642) made its maiden fight on 7 May 1969.

During the development programme the four original Sikorsky airframes were converted to Westland Sea King specifications and underwent a rigorous flight test programme. Westland replaced the original General Electric T58 engines with Rolls-Royce Gnome H 1400 turbo shaft engines and fitted a full-authority electronic engine control system and a Louis Newmark Mk31 Automatic Flight Control System (AFCS). Westland built the transmissions and rotor blades and installed the tactical equipment which included an Ecko AW 391 dorsal mounted search radar, Plessey 195 dunking sonar and other mission equipment. This would allow the Royal Navy Sea King HAS Mk1 to operate as a hunter-killer ASW helicopter in all weathers, day and night.

During the trials period Sea King XV370, as the lead aircraft, undertook the majority of early performance handling and AFCS work. Sea King XV371 was used primarily for handling and AFCS trials on the Louis Newmark Mk31 automatic flight control system. Sea King XV372 was used as the trial aircraft for the Rolls-Royce Gnome engine installation. On 15 January 1969 XV372 was severely damaged during a heavy landing on the Mendips after a power loss due to engine icing and was subsequently written off. Sea King XV373 HAS Mk1 was used for systems and avionic development and trialed the RN mission equipment including Plessey 195 sonar, radios and Ecko AW 391 search radar.

Westland immediately began a programme to update and improve the Sea King and exploit the aircraft's potential in the ASW, search and rescue and logistics support roles. Having gained several overseas customers for the ASW and search and rescue variant, Westland then introduced the Commando version to undertake the troop transport role. Westland Helicopters' Sea King improvement programme led to the introduction of more powerful engines, an uprated transmission system which helped increase payload and performance. The Sea King's all-up weight increased from 20,500lb to 21,500lb and combined with the latest developments in composite materials, avionics and mission systems, led to the latest Advanced Sea King variant. Capable of utilising a wide range of avionic suites, active and passive sensors and armed with advanced torpedoes or carrying any one of the entire range of air-to-surface missiles including Exocet and Sea Eagle, the Advanced Sea King is now capable of combining the Anti-Submarine Warfare (ASW) role with the Anti-Surface Vessel Warfare (ASVW) role making it a formidable hunter-killer.

The Royal Navy took advantage of the Sea King improvement programme and this resulted in a progression from the HAS Mk1 through HAS Mk2, Commando HC Mk4, the Mk2A (AEW), the HAS Mk5 and the latest deep dunking HAS Mk6. The introduction of the HAS Mk6 greatly increased the Royal Navy's ASW capabilities by making use of the latest digital technology. A new integrated digital processor and sonar allows the HAS Mk6 to update and display both active and passive data plus information from the MEL Sea Searcher radar and ESM equipment on a single CRT display screen increasing the Sea King's ability to find and destroy enemy submarines.

All the Royal Navy's Sea Kings have

now been fitted with Westland advanced composite rotor blades helping to increase performance and reduce operating costs even further. During 'Operation Granby' the Commando Sea King fleet, plus several Sea King HAS Mk5s were updated to the latest operational specification. These included the fitting of satellite navigation equipment and comprehensive defence aids suite all helping to increase the capabilities and survivability of the Sea King even further. Other updates should soon include fully Night Vision Goggle (NVG) compatible cockpit lighting and an Emergency Lubrication System for the main transmission.

1991 saw Royal Navy Sea Kings deployed to Saudi Arabia and the Gulf during 'Operation Granby' and to Kurdistan and Bangladesh for 'Operation Haven' and 'Operation Mana'. During July that year at RNAS Culdrose, the Sea King celebrated 21 years front-line service with the Royal Navy followed by the announcement that Westland and IBM had been selected to supply 44 EH101 'Merlin' helicopters to the Royal Navy as their next generation maritime ASW helicopter.

With over 300 Westland Sea Kings in service with ten countries around the world plus those built by Sikorsky, Agusta and Mitsubishi, the Sea King has secured its place in history as the most versatile, reliable and enduring maritime helicopter in the world.

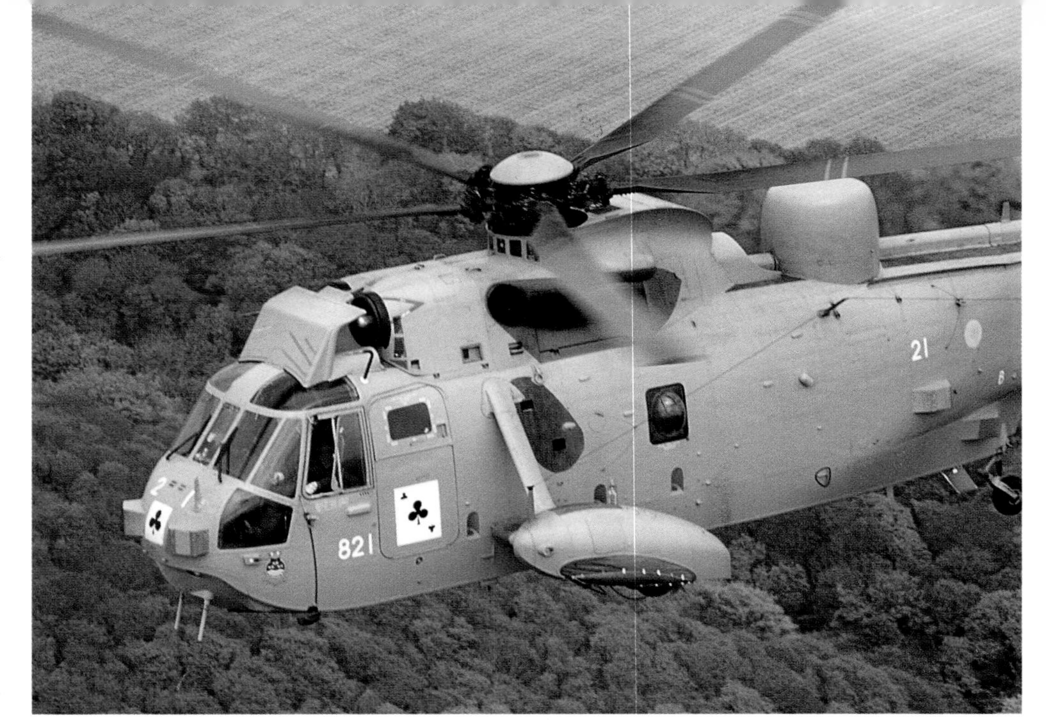

Centre right: Sea King XV370 was the original SH-3D designated S61D-2 to arrive in the UK and was registered G-ATYU. Used by Westland as their lead development aircraft, it was later transferred to the Empire Test Pilots' School at Boscombe Down. It retired in 1989. This photograph was taken at Broadlands 'Helimeet 89'. *(Patrick Allen)*

Top right: Sea King Mk 5 (XV705) was one of the first RN Sea King HAS Mk 1s to enter front-line service with 824 Squadron in 1970. It was later converted to a HAS Mk 2 then to a HAS Mk 5 operating with 706 Squadron. Today it is operated by 771 NAS in the SAR role and has been converted to a SAR Mk 5. This photograph was taken during the first month of Sea King operations by 771 NAS before the aircraft received its red painted nose and tail. *(Patrick Allen)*

Bottom right: 24 July 1991 and the Royal Navy celebrate 21 years front-line service with the Sea King with a line-up of seven variants and marks at RNAS Culdrose in Cornwall. The line-up includes the Westland 'Hack' Sea King Mk 2 (XZ570) fitted with EH101 Merlin mission equipment including the Blue Kestrel radar, Orange Reaper ESM and AQS903 sonics and 730 sonobuoy homer. The 'Hack' has flown over 400 hours on Merlin development since December 1986. Other Sea Kings in the line include an RN Sea King AEW 2A, HAS Mk 6, RN SAR Mk 5, RAF HAR Mk 3 and a Commando Mk 4. *(Patrick Allen)*

Below: Sea King HC Mk 4 (ZF115) was the first production Sea King to fly from the start with Westland advanced composite rotor blades on 3 June 1986. Most Royal Navy Sea Kings have now been retrofitted with composite main rotor blades which has increased performance and helped to reduce vibration etc. Operated by A&AEE Boscombe Down ZF115 is seen on its first flight at Westland. *(Westland)*

Below: Sea King XV642 was the first production HAS Mk 1 and spent its life as a trials and development aircraft. In 1970 XV642 went to Canada to undertake icing trials and was later used as the conversion aircraft for the HAS Mk 2. As a HAS Mk 2 XV642 was used as the Flight Testbed for Westland composite rotor blades. *(Westland)*

SECTION 2

WEST GERMAN NAVY SEA KING Mk41

The West German Navy ordered 22 Sea King Mk41s in June 1969. This was the first order for the successful SAR variant and was basically similar to the Royal Navy HAS Mk1 but, without the sonar equipment. The rear cabin was increased in size with the bulk head being moved aft by 5ft 8in and two additional bubble windows were fitted. The increased cabin size allowed room for up to 21 passengers and two crew, larger fuel tanks and a rescue winch were also included.

The first Sea King Mk41 (89+50) flew on 6 March 1972 and joined Marineflieger MFG.5, to undertake the SAR role based at Kiel-Holtenau. Like other export customers the West German Navy took advantage of the Royal Navy Foreign Training Unit at RNAS Culdrose and MFG.5 aircrew undertook their conversion training in Cornwall. During one of these training courses in January 1974 two West German Navy Sea King Mk41s (89+51, 89+55) along with three Royal Navy Sea King Mk1s from 706 Naval Air Squadron were involved in the dramatic rescue of the 480-ton Danish coaster *Merc Enterprise*. The Sea Kings snatched eleven survivors from the wreck in hurricane force winds and state 10 seas, 25 miles south of Plymouth. During the return flight Sea King Mk41 (89+55) callsign GERMANY 5, force landed in a field at Porthoustock in Cornwall. A US Army CH54 'TARHE' was sent over from its base in Germany to airlift the Sea King back to RNAS Culdrose.

The Mk41 production run lasted from 1972 through 1973 with the final Sea King Mk41 (89+71) flying on 21 August 1974. Sea King Mk41 (89+61) which first flew on 10 January 1974 was written off before delivery. This aircraft was replaced by Sea King Mk41 (WA830, 89+61) which first flew on 18 April 1975 and was delivered to the West German Navy on 23 July 1975.

Undertaking the SAR role with MFG.5 based at Kiel with detachments at Sylt, Borkum and Heligoland the West German Navy have undertaken a programme to update their Sea King fleet for the Anti-Surface Vessel (ASV) role. This was completed in 1988. This included the installation of the Ferranti Seaspray Mk3 radar for over-the-horizon targeting and allows the helicopter to be armed with the British Aerospace Sea Skua air-to-surface missile. The helicopters have also been fitted with the 'barn door' engine intake guards.

WEST GERMAN NAVY SEA KING Mk41

Sea King Mk41 (89+50) — 6/3/72 to Mk41 (89+51) — 26/6/72. 2 a/c.

Sea King Mk41 (89+52) — 7/6/73 to Mk41 (89+71) — 21/8/74. 20 a/c.

(Sea King Mk41 (WA765 — 89+61) written off)

Sea King Mk41 (89+61) — 18/4/75 second aircraft.

Below: The latest Sea Skua missile-capable German Navy Sea King Mk41 from MFG 5 based at Kiel, fitted with a nose-mounted Ferranti Sea Spray radar, barn door, chaff, flare and missile radar warning equipment. Now operating in both the SAR and anti-ship role MFG-5 Sea Kings operate from four locations along the North German coast. *(Patrick Allen)*

Below: A West German Navy Sea King Mk 41 (89+57) on a training mission. Like other NATO SAR units they undertake both military and civilian SAR missions. *(Westland)*

SECTION 3

INDIAN NAVY SEA KING Mk42

The Indian Navy purchased more Sea Kings from Westland than any other overseas customers. Since 1969 they have ordered 41 Sea Kings in four different variants including the latest Advanced Sea King.

In 1969 the Indian Navy placed an initial order for six Sea King Mk42s. These were in most respects identical to the Royal Navy HAS Mk1 and in 1974 Westland received a further order for an additional six Mk42s. The first Sea King Mk42 (IN501) flew on 14 October 1970 and these aircraft joined Nos 330 and 336 Squadrons in the ASW and training roles. In March 1980 three uprated Sea Kings, designated the Mk42A, were delivered. These aircraft, similar to the Royal Navy HAS Mk2, were fitted with a haul down capability to allow operations from small ships.

In July 1983 Westland received an order from the Indian Navy for 12 Advanced Sea King Mk42Bs with an option for eight more. Capable of operating in the ASW or ASUW roles the Advanced Sea King featured composite main and five-bladed tail rotors, an uprated transmission system and Rolls-Royce Gnome H 1400-1T turboshaft engines which increased the gross weight to 21,500lb. Mission equipment included the GEC Avionics AQS-902 sonobuoy processor and tactical processing system; MEL Super Searcher radar, Alcatel HS-12 dipping sonar, Chelton 7 homer and Marconi Hermies ESM system. In the ASV role the Mk42B can be armed with two Sea Eagle long-range anti-ship missiles or for the ASW role carry four Mk44/46 torpedoes or Mk11 depth charges. The first Advanced Sea King Mk42B (ZF526/IN513) flew on 17 May 1985 and remained as the development aircraft until 10 December 1990 when it was re-serialled IN532. The second build Sea King Mk42B (WA951/ZF527/IN514), flew on 18 July 1985 but was written off when it ditched into the Mediterranean during trials on 23 July 1987. As the aircraft crashed before delivery it was replaced by a second Sea King which had the same serial.

The Indian Navy also ordered a utility transport variant, the Sea King Mk42C. Six aircraft were ordered and like the Mk42B were powered by the uprated Rolls-Royce 1400-1T engines. Capable of carrying 28 troops in the utility role or 22 survivors plus a crew of four in the SAR role. The Mk42C had a similar avionics/navigation suite to the RAF Sea King HAR 3 although the MEL radar was replaced by a nose-mounted Bendix RDR 1400C radar. The first Sea King Mk42C flew on 25 September 1986.

ZF526/IN513 was the first Advanced Sea King Mk42B and first flew on 17 May 1985. Used as a trials aircraft it is seen here carrying a Sea Eagle missile. ZF526 was finally delivered to the Indian Navy on 10 December 1990. *(Westland)*

WESTLAND ADVANCED SEA KING Mk42B ASW/ASV

Crew of four in ASW role, 28 troops in utility role or 22 survivors in the SAR role.

Power Plant: Two Rolls-Royce Gnome H 1400-1T turboshafts rated at 1660shp.
Gross Weight: 21,500lb (9,752kg).
Empty: 13,000lb (5,896kg).
Max Payload: 8,000lb (3,628kg).
Range: 664nm (1,230km).
Weapons: Two Sea Eagle Missiles, Mk44/46 torpedoes, depth charges, rockets and machine guns.

The Advanced Sea King benefits from Westland advanced composite blade technology. The Mk42Bs are fitted with both main and tail composite rotor blades. The tail rotor of the Advanced Sea King has reverted back to the original five blades of the HAS Mk1. New tail rotor and blade design has increased blade performance even without the extra blade.

ANTI-SUBMARINE WARFARE

The Advanced Sea King can remain on station for two hours at 125nm (230km) carrying a realistic operational load.

ANTI-SURFACE UNIT WARFARE (ASUW)

The Advanced Sea King can carry a variety of air-to-surface missiles including Sea Eagle, Exocet AM39, Sea Skua, Harpoon, Otomat and Penguin (Mk42B — Sea Eagle). When operating in the ASV role the Advanced Sea King can remain on station for one hour whilst operating at 100nm (185km) from a land or ship base. The increased performance and sophistication of modern missiles has helped to increase stand-off distances, search radar data links, ESM and secure speech communications have also helped.

SEARCH & RESCUE

Capable of operating in all weathers, day and night, the Advanced Sea King has an eight-hour endurance with a typical radius of action of 310 (574km) with 10 survivors. The advanced AFCS allows accurate hover and hover trim facility at heights up to 140ft (43m).

INDIAN NAVY SEA KING Mk42

Sea King Mk42 (14/10/70) — IN501 to IN512 — 12 built
Sea King Mk42A (23/11/79) — IN551 to IN553 — 3 built
Sea King Mk42B (17/5/85) — IN513 to IN533 — 21 built
Sea King Mk42C (25/9/86) — IN555 to IN560 — 6 built

Top right: Advanced Sea King Mk 42B (IN518/ZG605) of the Indian Navy flying along the Dorset coast prior to delivery. *(Westland)*

Right: The Advanced Sea King including the Mk42B of the Indian Navy are capable of undertaking the ASW and ASUV roles. With their integrated mission equipment and weapon options they are amongst the most capable long range ASW/ASUV maritime helicopters available. *(Westland)*

SECTION 4

ROYAL NORWEGIAN AIR FORCE SEA KING Mk43

The Royal Norwegian Air Force were quick to see the potential of the Sea King in the SAR role. In 1970 they ordered ten Sea Kings which were designated Mk43s. These were basically similar to the West German Navy Mk41 and included the enlarged rear cabin allowing room for 21 passengers and two crew, extra bubble window and larger fuel tanks. Engines and dynamics were similar to the Royal Navy HAS Mk1 including the original five-blade tail rotor. In 1978 a single Sea King Mk43A (WA874/189) was ordered which included the uprated transmission and six-blade tail rotor of the HAS Mk2.

Operated by 330 Squadron based at RNoAF Bodo in Norway the Sea King Mk43 are used to provide military and civilian SAR cover along the hazardous Norwegian coastline and mountain areas. The first Sea King Mk43 (060) flew on 19 May 1972 and joined 330 Squadron on 16 December 1972. Norway is a demanding place to fly helicopters, particularly in the winter and even more so in the SAR role. Several of the original Sea King Mk43s have been written off (068 and 072) and two have been rebuilt after suffering accidents. Sea King Mk43 069 crashed into a frozen lake and was at first considered too badly damaged for repair. It was returned to Westland at Yeovil and after a major rebuild programme in 1989 it returned to 330 Squadron and carried out its first rescue just one week after arriving back at Bodo. Sea King Mk43 071 suffered an engine failure and ditched off Grimstead in 1991. The aircraft suffered structural damage during the recovery operation and was returned to Westland in October 1991 where it was repaired and brought up to Mk43B standard. In 1989 Westland received a multi-million pound contract to build one new Sea King Mk43B and to update the RNoAF Sea King fleet to the updated Mk43B standard. Improvements include a modernised navigational and radio suite and the installation of a nose-mounted weather radar and nose-mounted forward looking infra red (FLIR). The new build Sea King Mk43B was due to be delivered in July 1992 with Sea King 071 being the first aircraft to be converted to the new standard.

ROYAL NORWEGIAN AIR FORCE SEA KING Mk43

Sea King Mk43 — 060 — (19/5/72)
Sea King Mk43 — 062 — (21/6/72)
Sea King Mk43 — 066 — (30/6/72)
Sea King Mk43 — 068 — (18/7/72)
written-off 10/11/86
Sea King Mk43 — 069 — (30/7/72)
rebuilt 1989
Sea King Mk43 — 070 — (15/8/72)
Sea King Mk43 — 071 — (30/8/72)
rebuilt Mk43B — 1991/92
Sea King Mk43 — 072 — (9/9/72)
written-off 30/4/77
Sea King Mk43 — 073 — (21/9/72)
Sea King Mk43 — 074 — (30/9/72)
Sea King Mk43A — 187 — (6/7/78)
Sea King Mk43B — Serial ZH566 — 1992

Left: The latest RNoAF Sea King Mk41Bs lifting away from Westland Helicopters at Yeovil prior to their delivery in the summer of 1992. The nearest Sea King Mk43/071 first flew in August 1972 and was rebuilt to Mk43B standard after it crashed. Sea King Mk43B/322 is brand new and still retains its UK military registration ZH566.

The Mk43Bs are fitted with the latest night, adverse weather SAR equipment including composite rotor blades, improved navigation and communication systems, the latest MEL Sea Searcher colour radar, nose-mounted Bendix weather radar and Forward Looking Infra Red (FLIR). It is highly probable that the new SAR Sea Kings ordered by the RAF will include many of these features. *(Westland)*

Below: Sea King Mk43 (060) of 330 Squadron based at RNoAF Bodo was the first of ten Sea King Mk43s and first flew on 19 May 1972. Operating in the SAR role the Norwegian Air Force ordered a single Sea King Mk43A and Westland is now updating the fleet to the Mk43B standard. *(Westland)*

SECTION 5

PAKISTAN NAVY SEA KING Mk45

In December 1972 the Pakistan Navy ordered six Sea King Mk45s. These were to be used in the Anti-Submarine Warfare and SAR role and were identical to the Royal Navy HAS Mk1. The first Mk45 flew on 30 August 1974 with the other five following by the end of the year. Like the other foreign navy units the Pakistan Mk45s and their aircrew undertook their conversion training with the Royal Navy Foreign Training Unit at RNAS Culdrose.

During 1976 the Sea King Mk45s were converted to undertake the ASV role and were armed with the AM39 Exocet missile. Missile launching trials took place both in France and at Larkhill Ranges on Salisbury Plain between 1976/77. In 1986 a Sea King HAS Mk5 was converted to a Mk45 and flew on 9 April 1986. The aircraft was delivered on 11 January 1989.

Below: Pakistan Navy Sea King Mk45 (4514/G-BCNW) launching an Exocet AM39 missile during trials at Larkhill Ranges in 1976/77. The Mk45 can carry two Exocet missiles in the anti-shipping role. *(Westland)*

PAKISTAN NAVY SEA KING Mk45

Sea King Mk45 — (4510) — 30/8/74
Sea King Mk45 — (4511) — 10/9/74
Sea King Mk45 — (4512) — 17/9/74
Sea King Mk45 — (4513) — 30/9/74
written-off 8/2/86
Sea King Mk45 — (4514) — 27/11/74
Sea King Mk45 — (4515) — 10/12/74
Sea King Mk45B (ex-HAS Mk5, ZE421)
9/4/86

Below: Pakistan Navy Sea King Mk45 (4511) first flew on 10 September 1974 and was the second Mk45 to be built. The aircraft joined the RN Foreign Training Unit at RNAS Culdrose and was delivered to Pakistan in April 1977. *(Westland)*

SECTION 6

BELGIAN AIR FORCE SEA KING Mk48

On 22 April 1974 the Belgian Air Force ordered five Sea King Mk48s to undertake the SAR role. Westland had already had success in producing a SAR variant for several other overseas customers. These already included the Mk41 for the West German Navy and the Mk43 for the Royal Norwegian Air Force.

The Sea King Mk48 incorporated all the updates and improvements which had been installed in the Australian Navy Mk50 and Royal Navy HAS Mk2. These included the uprated Rolls-Royce 1600shp Gnome 1400-1 turboshaft engines, strengthened transmissions, six-blade tail rotor and enlarged rear cabin area. The cabin could accommodate up to 23 people or if configured in the Casualty Evacuation role, six stretchers and eight seats or three stretchers and 16 seats.

Operated by 40 Squadron Heli, based at Koksijde, Belgium, the Sea King role was to undertake the Belgian Air Force SAR commitment in conjunction with NATO and the principles laid down in the Convention of Chicago (1947) for the Belgian SAR region. This area extends throughout Belgium, Luxembourg and a section of the North Sea. The squadron is tasked from a Rescue Co-ordination Centre (RCC) based in Brussels and from two Rescue Sub-Centres (RSC) located at Koksijde and Luxembourg. They undertake both military and civilian tasks, day and night, 365 days a year.

The navigation and communication specifications for the Sea King Mk48 were extremely high. These have been regularly updated over the years and have also included the fleet being updated with new Westland composite main rotor blades. The aircraft are fully IFR-equipped having TACAN, VOR, DME, ILS, HF, VHF and UHF radios. Used in the day and night SAR role, the aircraft have a maximum take-off weight of 9.5 tons and an endurance of more than five hours. The AFCS allows automatic transition to the hover at heights of up to 43m (140ft) — clear of most ships' masts. The navigation suite and lightweight search radar allow efficient search patterns to be followed precisely. The AFCS also allows for Doppler-controlled hover, and an auxiliary hover trim gives the Winch Operator a 10 per cent over-ride control to adjust position of the helicopter and help facilitate rescue operations. The Belgian Air Force operate their Sea Kings with a crew of six. Two pilots, a navigator/SAROO (SAR Operations Officer), a flight engineer, a rescue diver and a doctor.

The first Sea King Mk48 (RS0-1) flew on 19 December 1975. The aircraft plus their 40 Squadron aircrews went down to RNAS Culdrose joining the Royal Navy Foreign Training Unit to undertake their conversion training. By the autumn of 1976 all five Sea Kings had arrived at Koksijde, Belgium, joining the existing fleet of Sikorsky S-58/HSS1s. The Sea King took over the SAR role completely in December 1978 and the last Sikorsky S-58 flew on 16 July 1986. The squadron continue to operate their five Sea King Mk48s in the SAR role plus three Alouette 111s.

As well as undertaking their SAR role the squadron is also tasked with the transportation of VIPs, trooping, para-jumping, load-lifting, military and civil medical evacuation, urgent transportation of vital organs, supporting police, gendarmerie, civilian and judicial authorities. They can also operate under the authority of the Naval Command at Zeebrugge. During heavy weather, when normal pilot service for ships inbound to Flushing is impossible, the squadron will assist.

On average the squadron Sea Kings under-

Right: Belgian Air Force Sea King Mk 48 (RS03) first flew in April 1976 and is seen here undertaking a SAR training mission. All five Sea King Mk 48s operate with 40 Squadron based at Koksijde and are kept in superb condition. *(Westland)*

take 75 SAR missions annually, involving a variety of rescues both on land and at sea. In 1990 a total of 68 people were rescued from a variety of situations including the following:

26/ 1/1990: The Russian factory ship *Briz* capsizes in the Terschelling area (Northern Netherlands) in storm force winds. Nineteen people are brought ashore.
15/ 2/1990: Five people rescued from a drifting oil platform.
4/10/1990: Four hikers and two dogs are rescued when they become cut off by the tide.
10/10/1990: After an explosion aboard the *Chioz Faith* the cargo catches fire. Three crew members are rescued.

BELGIAN AIR FORCE SEA KING Mk48
5 a/c

Sea King Mk48 — RS01 (19/12/75)
Sea King Mk48 — RS02 (8/ 4/76)
Sea King Mk48 — RS03 (12/ 4/76)
Sea King Mk48 — RS04 (7/ 5/76)
Sea King Mk48 — RS05 (7/ 6/76)

Top right: The Belgian SAR Sea King crews normally comprise two pilots, a navigator/ SAROO (SAR operations officer) who also operates the radar, a flight engineer, a rescue diver and a doctor. *(Patrick Allen)*

Bottom right: A Sea King Mk 48 (RS03) of 40 Squadron based at Koksijde is seen flying along the Belgian coast in 1987. The Belgian Air Force have, over the years, continually updated their five SAR Sea King Mk 48s with the original metal blades being replaced by composite blades and more recently a navigation systems update has been completed. *(Westland)*

Below: Belgian Sea King Mk 48s RS01, RS02 and RS03 seen in close formation at their home base during 40 Squadron celebrations to commemorate 30 years of SAR on 6 July 1991. *(Patrick Allen)*

SECTION 7

ROYAL AUSTRALIAN NAVY SEA KING Mk50/50A

The Royal Australian Navy (RAN) ordered ten Westland Sea King Mk50s in 1972. Twelve aircraft were eventually built with two Mk50As being built as attrition replacements in 1982.

The Mk50 took advantage of the continued development by Westland in the performance and payload of the Sea King. These improvements eventually led to the Royal Navy updating their existing HAS Mk1 fleet to the more powerful HAS Mk2,

The Mk50 was equipped with the uprated 1660shp Rolls-Royce Gnome H1400-1 engines, strengthened transmission and a six-blade tail rotor. These improvements brought the all-up weight of the Sea King Mk50 to 21,000lb. Operated by HS817 Squadron based at HMAS *Albatross*, the Sea Kings are used primarily for the ASW role. They are also tasked with Search and Rescue, VIP, Mine Clearance and Diverless Helicopter Weapons Recovery Systems (DHWRS). In their ASW role two torpedoes are carried and when configured for utility operations 13 passengers or freight can be carried internally and up to 2,700kg is carried underslung. In their SAR role the winch can lift 270kg from 240 feet below the aircraft and the Ecko 391 radar is used for navigation, surface surveillance, ASW and SAR. The RAN Sea Kings are similar in most respects to the Royal Navy HAS Mk5 except for their sonar equipment. The Australian Sea Kings are fitted with the American AQS 13B dipping sonar which allows them to work with RAAF Orions and surface units. The AQS 13B has over 500ft of sonar cable which gives it a 100ft advantage over the UK version. Over the years the Sea Kings have been steadily updated with the original radar being replaced with the more powerful MEL Sea Searcher and improved Doppler TANS.

The first Sea King Mk50 (N16-098) flew on 30 June 1974 and in October 1974 the first RAN Sea King Flight was formed at the Foreign Training Unit, RNAS Culdrose, Cornwall. HS817 Squadron spent a year working-up on the new aircraft. On 2 February 1976 HS817 reformed back in Australia with ten Mk50 Sea Kings. As an ASW Squadron HS817 has deployed in HMA Ships *Melbourne*, *Stalwart*, *Success* and *Tobruk* undertaking their various roles which have included supporting various exercises in Australia, South East Asia, China and Hawaii for RIMPAC exercises.

Today seven Sea Kings remain in service (five Sea King Mk50s and two Sea King Mk50As) with HS817 Squadron. The squadron regularly produces five serviceable aircraft on the flight line which are maintained by a team of 90 aircraft technicians. The squadron has nine aircraft crews and each crew consists of two pilots, an observer (tactical navigator) and an aircrewman (sonar and winch operator). RAN Sea Kings are expected to remain in service well into the nineties.

Below: One of the five remaining Westland Sea King Mk 50s of 817 Squadron which was formed in the UK in 1974 before becoming operational at HMAS *Albatross* on 2 February 1976. *(RAN)*

WESTLAND SEA KING Mk50

Sea King Mk50 — N16-098 (30/6/74) written-off (24/5/79)
Sea King Mk50 — N16-100 (6/8/74)
Sea King Mk50 — N16-112 (16/10/74) written-off (13/7/86)
Sea King Mk50 — N16-113 (24/10/74) written-off (13/11/76)
Sea King Mk50 — N16-114 (13/11/74)
Sea King Mk50 — N16-117 (13/12/74)
Sea King Mk50 — N16-118 (3/1/75) written-off (21/9/75)
Sea King Mk50 — N16-119 (4/2/75) written-off (9/5/77)
Sea King Mk50 — N16-124 (25/3/75)
Sea King Mk50 — N16-125 (8/5/75)

SEA KING Mk50A

Sea King Mk50A — N16-238 (7/12/82)
Sea King Mk50A — N16-239 (9/ 2/83)

Below: Two RAN Sea King Mk 50s of 817 Squadron deck landing on an RN ship watched by an 846 Squadron Commando Wessex HU Mk 5. *(Westland)*

SECTION 8

EGYPTIAN AIR FORCE SEA KING COMMANDO Mk1 and 2

Five Sea King Commando Mk1s were ordered in 1973 by Saudi Arabia on behalf of the Egyptian Air Force. These were essentially similar to the Sea King Mk41. Instead of the larger cabin area being used for the SAR role they were used in the utility role moving troops and cargo. The first Sea King Commando Mk1 (262) flew on 12 September 1973.

Westland were quick to see the potential of the Sea King in the utility troop transport role and with the potential of overseas sales began a detailed programme re-designing the aircraft. This included the fitting of the uprated Rolls-Royce Gnome H1400-1 engines, transmission and six-blade tail rotor of the HAS Mk2 and Mk50. The Westland designers also modified the Sea King's landing gear. They removed the sponsons and fitted a non-retractable undercarriage which was attached to short stub wings. Other changes included the removal of the main rotor blade fold capability and the fitting of a basic avionics suite. The new Commando Mk2 could carry up to 28 fully equipped troops or 2,720kg (6,000lb) of cargo or 8,000lb underslung.

In 1974 Saudi Arabia on behalf of the Egyptian Air Force ordered 17 Sea King Commando Mk2s and two VIP Mk2Bs. One of the Egyptian Air Force requirements was for the installation of sand filters for the Sea King Commando Mk2 fleet. The sand filters replaced the existing 'barn door' in front of the engine. The filter comprised twin fans that used centrifugal force to blow incoming sand out sideways from the engine inlets and into collector boxes. The sand could then be sucked out and away by an electric fan. The filters proved excellent at helping to keep

Below: Two Egyptian Air Force Sea King Commando Mk2s with British civil registrations prior to delivery. *(Westland)*

sand damage to the engines at a minimum. The Royal Navy Commando Sea King HC Mk4s later benefited from this early work when their helicopters were amongst the few Allied helicopters that possessed effective sand filters at the beginning of the Gulf War. The first Egyptian Air Force Sea King Commando Mk2 (721) flew on 16 January 1975.

During 1978 the Egyptian Air Force ordered a further four Commando Sea Kings to be modified for the Electronic Counter Measures and Electronic Surveillance roles. Designated the Commando Mk2E the first of four aircraft (SU-BBJ) fitted with Elettronica IHS-6 ECM/ESM equipment flew on 1 September 1978.

EGYPTIAN AIR FORCE SEA KING COMMANDO

Sea King Commando Mk1	(12/9/73)	5 a/c
Sea King Commando Mk2	(16/1/75)	17 a/c
Sea King Commando Mk2B	(13/3/75)	2 a/c
Sea King Commando Mk2E	(1/9/78)	4 a/c

EGYPTIAN NAVY SEA KING Mk47

Having placed an order for the Sea King Commando, Saudi Arabia placed an order for six ASW Sea Kings on behalf of the Egyptian Navy. Designated the Sea King Mk47 they were essentially similar to the Royal Navy HAS Mk2 and Australian Navy Mk50. Like the other overseas navies who had purchased their Sea Kings from Westland, the Egyptian Navy Sea King Mk47s and their aircrew undertook conversion training at the Royal Navy Foreign Training Unit at RNAS Culdrose.

The first Sea King Mk47 (WA822) flew on 11 July 1975 and the final aircraft flew that November.

SEA KING Mk47 — 6 a/c

Sea King Mk47 (WA822)	— 11/ 7/75
Sea King Mk47 (WA823)	— 7/ 8/75
Sea King Mk47 (WA824)	— 4/ 9/75
Sea King Mk47 (WA825)	— 25/ 9/75
Sea King Mk47 (WA826)	— 23/10/75
Sea King Mk47 (WA827)	— 5/11/75

Top right: The first Egyptian Commando Mk2E (WA866) equipped for electronic counter-measures operations first flew on 1 August 1978. Westland built four Mk2Es in 1978. *(Westland)*

Right: Egyptian Air Force Sea King Commando Mk2B VIP version showing the additional cabin window and pre-delivery registration. *(Westland)*

SECTION 9

QATAR EMIRI AIR FORCE SEA KING COMMANDO Mk2A, Mk2C and Mk3

The success of the Sea King uprating programme and the development of the Commando Sea King Mk2 for the Egyptian Air Force led to an order by the Qatar Emiri Air Force for the Commando Sea King to be built in three versions.

In 1974 the Qatar Air Force ordered four Commando Sea Kings in two variants. Three of these Sea Kings were destined for the general utility role and were designated the Mk2A. The remaining Sea King was designated the Mk2C and was fitted out for the VIP role. All were powered by Rolls-Royce Gnome H1400-1 engines and had the six-blade tail rotor and improved transmission system. The first Sea King Commando Mk2A (QA20) flew on 9 August 1975 followed by the Mk2C VIP version on 9 October 1975. The final two Mk2As (QA21 and QA22) flew on 10 March 1976 and 16 March 1976 respectively.

A second order for eight uprated Commando Sea Kings was received and these were designated the Commando Mk3. The Mk3 aircraft were fitted with the more powerful Rolls-Royce Gnome-1400-1T turbo-shaft engines which developed 1,600shp and increased the AUW to 21,500lb. The Commando Mk3 was to undertake the general utility role but would also carry a variety of weapons including two AM39 Exocet missiles for the anti-shipping role. The first Qatar Commando Sea King Mk3 (QA30) flew on 14 June 1982 followed by the remaining seven aircraft with the last one (QA37) flying on 5 October 1983.

QATAR AIR FORCE COMMANDO SEA KINGS

Sea King Commando Mk2A — 3 a/c
Sea King Commando Mk2C — 1 a/c
Sea King Commando Mk3 — 8 a/c

Top right: Qatar Air Force Sea King Commando Mk 3 (QA30) armed with two AM39 Exocet missiles. QA30 first flew on 14 June 1982 and was delivered in November. *(Westland)*

Right: Qatar Air Force Commando Mk 2 (QA20) first flew on 21 January 1976 and was delivered in February. This photograph shows the Sea King fitted with its sand filter. *(Westland)*

SECTION 10

MINISTRY OF DEFENCE PROCUREMENT EXECUTIVE (PE)

DRA FARNBOROUGH and BEDFORD AEROPLANE AND ARMAMENT EVALUATION ESTABLISHMENT BOSCOMBE DOWN

EMPIRE TEST PILOTS' SCHOOL BOSCOMBE DOWN

Over the years the Sea King has been the subject of numerous trials and developments and these continue to this day. All military aircraft and helicopters undergo numerous pre-service trials. These assess new aircraft, modifications to existing aircraft, newly integrated systems or new weapons etc. The responsibility of assessing the potential of new aircraft and their systems including new technology is the responsibility of A&AEE Boscombe Down. These tests and assessments can begin at the earliest stages of development and the Establishment is closely involved with the manufacturers in the fields of test methods and instrumentation, systems appraisal, rig tests and during suitable stages in the manufacturers' flight trials undertaking preliminary assessments of new aircraft. The major role of the Establishment however, is the full testing and evaluation of new aircraft, systems or weapons on behalf of the Controller Aircraft leading to a CA release and to confirm that they are reliable, safe and possess the handling characteristics appropriate to the aircraft's role.

Aircraft at the Defence Research Agency (DRA) at Farnborough and Bedford undertake more pure experimentation. This involves both future aircraft and their systems and includes cockpit technology. The DRA working closely with industry assesses the potential of these state-of-the-art systems. In recent years this has included designing and assessing future flight control systems and cockpits such as the 'glass' cockpits using CRTs, fly-by-wire and fly-by-light controls and advanced radar systems etc. Both Farnborough and Bedford have their own Sea King Mk4Xs (ZB506 and ZB507) which were delivered in 1982/83. These are basically standard Sea King HC Mk4s fitted with the dorsal-mounted Sea Searcher radomes. Both aircraft are used as development aircraft and are modified for use as flight test vehicles.

Right: Sikorsky Sea King SH-3D XV370 was the first Sea King brought into the UK by Westland in 1966 and later used by the ETPS at Boscombe Down. *(Patrick Allen)*

A&AEE BOSCOMBE DOWN/ETPS

There are several Divisions within A&AEE Boscombe Down which include a Flying Division of which the Empire Test Pilots' School is one of the squadrons.

These divisions include:

ARMAMENT DIVISION

This Division is responsible for assessing the safe and reliable carriage, release and jettison of weapons from aircraft and helicopters. This includes flight trials and the comprehensive analysis of data.

ENGINEERING DIVISION

Responsible for the safety and performance of the engineering and computer systems of aircraft throughout the climatic operating range. This includes analysing data from flight test programmes and electromagnetic compatibility assessments. The Division is also responsible for the development and clearance of paratrooping and supply dropping methods and equipment.

NAVIGATION AND RADIO DIVISION

This is the largest assessment division of A&AEE and is responsible for the evaluation of avionic equipment and other airborne radio systems. This includes navigation and weapon delivery accuracy measurements, radar, communications, reconnaissance and electronic warfare systems. Trials are carried out either prior to the introduction of a new aircraft to the Service or as existing aircraft are upgraded with new equipment to extend their operational capabilities.

SEA KING HC Mk4 (ZF115)

Modern military aircraft navigation and communication systems represent the forefront of aerospace technology. The Division manages two laboratory aircraft including a Sea King HC Mk4 (ZF115). This is operated by the Rotary Wing Test Squadron, and is used for testing and recording navigation and communication systems and performance over the full range of aircraft operating dynamics.

In 1986 Sea King HC Mk4 (ZF115) replaced Sea King XV373 which was one of the original Sikorsky SH3D Mk1s brought over by Westland from the United States as a development aircraft. Used as a laboratory aircraft the Sea King HC Mk4 is comprehensively instrumented to determine the performance of systems under test and carries a state-of-the-art navigation system which acts as the datum for any equipment under trial. The aircraft is also fitted with a digital recording system which includes the use of a data bus, plus a set of calibrated aerials for radio systems performance testing.

The Division also operates the Radio Trials Centre facility to measure and calibrate communications and radio navigation equipment performance.

Below: Sea King ZF115 is used by Boscombe Down as their Navigation and Radio Division flying test bed. ZF115 was the 30th Commando Sea King HC Mk 4 and the first production Sea King to fly from the start with composite main rotor blades on 3 June 1986. Almost all RN front-line Sea Kings have now been retrofitted with Westland composite main rotor blades. *(A&AEE Boscombe Down)*

PERFORMANCE AND TRIALS MANAGEMENT DIVISION

This Division is split into three groups: Performance, Trials Management and Computer and Instrumentation Groups. They specialise in the assessment of flying qualities, air vehicle and engine performance, engine handling and cockpit aspects. They plan, control and co-ordinate trials efforts and provide the main data handling and computer resources for A&AEE. They also design and calibrate trials instrumentation and are the leading authority in the development of instrumentation and software to meet new and often demanding trials requirements.

FLYING DIVISION & EMPIRE TEST PILOTS' SCHOOL

The Flying Division is responsible for operating all aircraft engaged on Boscombe Down trials. The Division has two test squadrons — Fixed Wing Test Squadron and Rotary Wing Test Squadron. Both Squadrons work in close liaison with the Assessment Divisions' trials officers, the squadrons' air-crew provide the operational experience necessary for valid assessments. All military aircraft destined for use by the Services pass through the hands of aircrew from these test squadrons.

The Test Squadrons are manned by air-crew of all three Services with appropriate operational backgrounds. Most of the pilots are also graduates of the Empire Test Pilots' School or equivalent schools in France and the United States.

The Rotary Wing Test Squadron has a total of 10 test pilots from all three Services. Their busy programme includes the flight testing of new aircraft leading to their CA Release and the trialing of new equipment, systems and weapons. They have recently been involved in test flying the latest Royal Navy Sea King HAS Mk6 as well as flight testing the numerous operational enhancements fitted to the Sea King during 'Operation Granby'. The Squadron also undertake numerous operational flight trials on both new and existing helicopters evaluating recently fitted mission systems. These have included deck landing trials on various ships, underslung load, night vision goggle compatibility and aircraft icing trials.

EMPIRE TEST PILOTS' SCHOOL (ETPS)

The School is an integral part of the Flying Division, although it holds its own Charter. The School runs courses from February to December each year with an annual intake of about 20 student test pilots and engineers, up to half of whom may be from overseas. The aim of the School is to provide test pilots for test flying duties in aeronautical research and development establishments within the Service and industry.

ROTARY-WING COURSE

Whilst based at Farnborough in 1963, the School set up its first Rotary-Wing Course. This was to meet the ever-increasing need for professional helicopter test pilots. The School moved back to Boscombe Down in 1968 where it remains to this day. The Rotary-Wing Course has an intake of eight students per year which includes two engineers.

Below: Sea King HC Mk 4 (ZG829) first flew on 10 April 1989 and was delivered to Boscombe Down EPTS on 3 May 1989. By November 1991 the helicopter had flown 270 hours with the EPTS. *(A&AEE Boscombe Down)*

The ETPS has several of its own helicopters including a Sea King HC Mk4 (ZG829). This aircraft has been extensively fitted out with a variety of flight test instrumentation for the collection and storage of Flight Data. This includes a Nose Probe Actuator for measuring the angle of attack of the airflow coming onto the aircraft, an instrumented cargo hook and an onboard airborne performance computer. This measures various aircraft parameters including aircraft weight, performance and handling for specific climatic conditions and temperatures. There is a rear cabin console which is used by the Flight Test Engineer students for their evaluations.

Students must learn to assess the handling qualities and systems of their aircraft for specific roles and missions. They must analyse concisely the ability of the aircraft or system to perform the particular task in question. This investigation must be presented later either orally or in writing and the School places great emphasis on the presentation of these reports by its students.

The Sea King is used by the School for large aircraft handling evaluations and as the systems aircraft which includes navigation and radio equipment. The Sea King should soon be fitted with an Omega and Satellite (GPS) Navigation suite plus R-NAV2. Also to be fitted is a nose-mounted Bendix 1400C weather radar making the Sea King and its systems a demanding aircraft to evaluate. The flying programme includes cockpit assessments, low speed manoeuvres, engine and rotor governing, ceiling climbs, flight envelope and manoeuvre boundaries, autorotations and engine-off landings, stability and control reviews, systems and instrument flying assessments.

During their course students undertake around 150 flight hours in the School's helicopters which include a Scout, Lynx, two Gazelles and the Sea King. Approximately 30 hours are spent flying the Sea King.

DRA at BEDFORD AND FARNBOROUGH

Sea King HC Mk4X — ZB506 (19/11/82)
Sea King HC Mk4X —ZB507 (10/ 1/83)

A&AEE BOSCOMBE DOWN

Sea King HC Mk4 —ZF115 (3/6/86)
(Navigation & Radio Division)

ETPS BOSCOMBE DOWN

Sea King HC Mk4 — ZG829 (10/4/89)
w/off 20/10/92

Below: Boscombe Down Sea King line-up showing five Variants of the Sea King.

XV373 was one of the original Sikorsky SH-3Ds used by Westland for their Sea King development programme and used by them as the avionics and systems aircraft. It was passed over to A&AEE Boscombe Down for Navigation and Radio trials and was withdrawn from service in 1986. Sea King XV373 was replaced by Sea King HC Mk4 (ZF115) in 1986. Other aircraft in the line-up include:

Royal Navy Sea King HAS MK2.
RAF Sea King HAR3.
Royal Navy Commando Sea King HC Mk4.
Royal Navy Sea King HAS Mk5.

(A&AEE Boscombe Down)

SECTION 11

ROYAL NAVY SEA KING HAS Mk1

Having ordered 56 Sea King HAS Mk1s on 27 June 1966, the first production Royal Navy HAS Mk1 (XV642) flew on 7 May 1969 and spent the remainder of its life in the development and trials programme. In 1970 it undertook icing clearance trials in Canada and was later modified to HAS Mk2 and became the development aircraft for the Westland advanced composite main rotor blade programme. On 11 August 1969 Commander V. Sirett, RN, the Commanding Officer of 700S Squadron flew XV645 to RNAS Culdrose to form the Intensive Flying Trials Unit (IFTU). Equipped with six Sea King HAS Mk1s (XV644, XV645, XV646, XV647, XV648, XV649), 700S Squadron was commissioned at RNAS Culdrose on 19 August 1969. Between August 1969 and the time the unit disbanded in May 1970 the IFTU had flown over 1,700 hours with each aircraft averaging around 468 hours as the Unit evaluated the anti-submarine role of the new Sea King.

On 24 February 1970 the Royal Navy Sea King HAS Mk1 went into front line service when 824 Naval Air Squadron was commissioned at RNAS Culdrose. In June of that year, 824 Squadron embarked in HMS *Ark Royal* and the Royal Navy had their first fully effective hunter-killer helicopter with an all-weather day and night capability, operating in the front line. By March 1973 seven Royal Navy squadrons were equipped with the Sea King HAS Mk1.

- 706 Naval Air Squadron, RNAS Culdrose (Sea King ASW Advanced Training and Conversion).
- 737 Naval Air Squadron, RNAS Portland (Training for ships flight role).
- 814 Naval Air Squadron, HMS *Gannet*, Prestwick (HMS *Hermes*).
- 819 Naval Air Squadron, HMS *Gannet*, Prestwick (RFA flights and support for Clyde-based submarines).
- 820 Naval Air Squadron, RNAS Culdrose (HMS *Blake*).
- 824 Naval Air Squadron, RNAS Culdrose (HMS *Ark Royal*).
- 826 Naval Air Squadron, RNAS Culdrose (Large fleet carriers and cruisers, i.e. HMS *Eagle*).

SEA KING HAS Mk1 SPECIFICATION

Built by Westland for the Royal Navy, the Sea King had twice the capability of the Wessex 3, flying sorties of four hours, covering a search area four times greater. New features included an automatic power-folding five-blade main rotor, retractable undercarriage and boat-type hull with sponsons. Mission equipment included the Echo A391 search radar, Plessey 196 medium frequency sonar, Marconi Doppler navigation system and a Louis Newmark Mk31 Automatic Flight Control System (AFCS) which allowed the aircraft to be flown into and out of a hover, over the sea at night.

The first production Sea King HAS Mk1 flew on 7 May 1969 and the Royal Navy bought 56 aircraft in the first production contract.

The first front line Sea King HAS Mk1 squadron was 824 Naval Air Squadron commissioned at RNAS Culdrose on 24 February 1970 and embarked in HMS *Ark Royal*.

The second front line squadron to be equipped with the Sea King HAS Mk1 was 826 Naval Air Squadron which was commissioned on 2 June 1970 at RNAS Culdrose and embarked in HMS *Eagle*.

SEA KING HAS Mk1s — 56 Built

XV642 (7/5/69) to XV677 (18/12/70) (36 a/c)
XV695 (2/1/71) to XV714 (15/ 5/72) (20 a/c)

SEA KING HAS Mk1

Crew: 4 (two pilots, one observer, one sonar operator).
Powerplant: two 1,500shp Rolls-Royce Gnome H1400 gas turbine engines.
Dimensions: Rotor diameter 62ft. Length 55ft 9.75in.
Weight: Empty 12,170lb. Loaded 20,500lb.
Performance: Max speed 161mph. Climb 3,000ft/m. Ceiling 14,700ft. Range 598 miles.
Armament: 4 × Mk44 homing torpedoes.

Below: Sea King HAS Mk 1 from 826 Squadron with the Cornish coast behind. 826 Squadron received their Sea Kings in June 1970. XV663 was the 22nd Sea King to be built and first flew on 15 April 1970. It was later converted to a HAS Mk 2 then to a HAS Mk 5 and is flying today as a HAS Mk 6. *(Westland)*

Below: A flypast of St Michael's Mount in Cornwall in 1975/76 by Sea Kings of the Royal Navy Foreign Training Unit (RNFTU) based at RNAS Culdrose. The RNFTU was commissioned at RNAS Culdrose on 12 June 1973 and undertook Sea King conversion training for overseas customers of the Sea Kings. Aircraft shown in this photograph include an Egyptian Mk 47, Pakistan Mk 45, West German Navy Mk 41 and a Royal Navy HAS Mk 1. *(Westland)*

SECTION 12

ROYAL NAVY SEA KING HAS Mk2

Almost immediately the Sea King entered Fleet Air Arm service, they began to secure a place in history as one of the most versatile, reliable and capable maritime helicopters in the world. The Sea King not only excelled in the ASW role, but also provided the Royal Navy with an outstanding long-range Search and Rescue (SAR) helicopter. In 1970 a Sea King HAS Mk1 established the longest recorded non-stop flight by a British services helicopter flying from Land's End to John o' Groats. The distance covered was 602.95 miles with an average speed of 139.49mph. 1970 also saw a RNAS Culdrose-based Sea King HAS Mk1 undertaking a record long-range SAR mission on 14 May 200 miles into the Atlantic. This was the beginning of a long list of rescues as the Sea King established a place for itself as a highly capable SAR helicopter. During 1971, two 824 Naval Air Squadron Sea Kings broke more records, flying a 430-mile non-stop SAR mission in gale force Atlantic conditions, to assist the Norwegian ship *Anatina*. In October of that year in the South China Seas, Sea King HAS Mk1s from 826 Naval Air Squadron rescued 40 crew members from the American ship *Steel Vendor*.

The success of the Westland-built Sea King led to several orders from overseas customers for their own Sea King variants. These included the Mk41 SAR version for the West German Navy, the Mk42 for the Indian Navy, the Mk43 SAR variant for the Royal Norwegian Air Force, a Mk1 Commando version for the Egyptian Air Force, the Mk45 for the Pakistan Navy and the Sea King HAS Mk50 for the Royal Australian Navy. Westland progressed with their update programme for the Sea King, improving both payload and performance. The Australian Navy order allowed Westland to forge ahead with this programme and to produce a more powerful version, which would also benefit other customers, Egypt in particular, who required their helicopters to operate in hot climatic conditions. The Australian Mk50 was fitted with uprated Rolls-Royce 1,600shp Gnome H1400-1 gas turbine engines and a strengthened transmission system including main, intermediate and tail rotor gearboxes to cope with the additional power. Also added was a sixth tail rotor blade allowing better hover performance at the aircraft's increased all-up weight.

The Royal Navy continued their success with the Sea King HAS Mk1. In 1971 the first Sea King Simulator arrived at RNAS Culdrose and in June 1972 the final Sea King HAS Mk1 was delivered. The success of the Westland Sea King was such that on 19 June 1973 the Royal Navy Sea King Foreign Training Unit was commissioned at RNAS Culdrose to provide conversion and operational training to the foreign navies who had purchased the Westland Sea King. They included the German Navy Mk41s, Pakistan Navy Mk45s and the Egyptian Mk47s. In January 1974 Sea Kings from RNAS Culdrose including German Navy Mk41s operating with the RNFTU rescued 11 men from the 480-ton Danish Coaster *Merc Enterprise* which had capsized in hurricane force winds and state 10 seas, 25 miles south of Plymouth.

With the ever increasing threat posed by the Soviet Navy the Royal Navy decided to take advantage of the improved capability of the uprated Sea King and in November 1974 the Royal Navy placed an initial order for 13 new Sea King HAS Mk2s which was later increased to 21. A programme to convert the existing fleet of HAS Mk1s was undertaken by both Westland and the Royal Navy at RNAY Fleetland.

The First Flight of a production Sea King HAS Mk2 (XZ570) took place on 18 June 1976 with 826 Naval Air Squadron re-

Right: Sea King HAS Mk 2 on a 'Dunker' training mission with 706 Naval Air Squadron. Tasked with the training and conversion of Royal Navy ASW Sea King crews. *(Westland)*

equipping with the HAS Mk2 in December 1976. Eventually, all Royal Navy front line ASW Sea King Squadrons converted to the HAS Mk2 including 706 Naval Air Squadron, the RNAS Culdrose-based training and conversion Squadron. The Royal Navy Sea King HAS Mk2 introduced uprated Rolls-Royce 1,600shp Gnome 1400-1 engines, six-bladed tail rotor and strengthened transmission system. Also fitted was the 'barn door' engine intake FOD guard with fluid de-icing strips which is now standard on all Royal Navy Sea Kings.

The first production Sea King HAS Mk2 (XV570) was retained by Westland and used as the prototype and development aircraft. It was later used as a development/trials aircraft in the Sea King HAS Mk5 programme and helped in the continued development of Westland composite rotor blades. The aircraft has more recently been the flight test aircraft for the EH101 Merlin mission equipment system and has been fitted with the Ferranti 'Blue Kestrel' radar.

By 1980 all of the original Sea King HAS Mk1s had disappeared from the Fleet Air Arm inventory. The additional performance of the Sea King HAS Mk2 was put to good use as the helicopter undertook numerous operational and SAR missions. In February 1978 RNAS Culdrose Sea Kings performed one of the most dramatic air/sea rescues of all time, hovering within inches of the towering legs of the oil rig *Orion*. The crews were decorated for their precision flying and hovering skills. In 1979 Royal Navy Sea Kings were again put to good use as they assisted in the Fastnet Race disaster.

The Sea King HAS Mk2 saw action during 'Operation Corporate' when 825 Naval Air Squadron was reformed in May 1982 and embarked in Royal Fleet Auxiliary ships for the Falklands.

The Sea King HAS Mk2s had their ASW equipment removed allowing room for up to 16 stretchers or 24 troops. The squadron was tasked for Logistics Lift, Trooping and Casevac duties. They saw action both at sea and during the retaking of the islands. 825 Naval Air Squadron Sea Kings undertook numerous casevac missions and took part in the *Sir Galahad* rescue. The squadron disbanded in September 1982 after an eventful five months.

SEA KING HAS Mk2 SQUADRONS

706 Naval Air Squadron, RNAS Culdrose.
814 Naval Air Squadron, HMS *Gannet*, Prestwick.
819 Naval Air Squadron, HMS *Gannet*, Prestwick.
820 Naval Air Squadron, RNAS Culdrose.
824 Naval Air Squadron, RNAS Culdrose.
825 Naval Air Squadron, RNAS Culdrose.
826 Naval Air Squadron, RNAS Culdrose.

WESTLAND SEA KING HAS Mk2s (21a/c)

XZ570 (18/6/76) to XZ582 (13/7/77) — 13 a/c (1st flt)
XZ915 (7/7/79) to XZ922 (24/8/79) — 8 a/c

Below: Five Westland Sea King HAS 2s from 824 NAS seen in formation passing the Penlee lifeboat house. 824 Squadron was the first front line Sea King squadron commissioned in February 1970 embarking onto HMS *Ark Royal*. The squadron re-equipped with the Sea King HAS Mk 2 in 1977 and became the parent unit for RFA Sea King flights. The squadron's 'Speedbird' emblem is clearly shown in this photograph. *(Westland)*

SECTION 13

ROYAL NAVY SEA KING AEW Mk2A

Born out of conflict the Sea King AEW Mk2A was designed and developed within a record 11 weeks. The lack of an adequate Airborne Early Warning (AEW) capability proved disastrous to the Royal Navy during the Falklands War. Using ship's radar to locate and classify low flying Argentine aircraft or sea skimming missiles, left little time to mount an effective defence. To counter this, Westland using two Sea King HAS Mk2 aircraft (XV650/XV704) mounted and integrated a Search Water radar which was housed externally, in a retractable and inflatable radome, on the starboard side of the aircraft.

The design proved an outstanding success and by August 1982 both modified aircraft were deployed aboard HMS *Illustrious* assigned to 'D' Flight, 824 Naval Air Squadron and were heading for the South Atlantic. On 9 November 1984, 849 Naval Air Squadron was commissioned at RNAS Culdrose and became the Fleet Air Arm's Sea King AEW squadron. A total of nine Sea Kings have been converted to the AEW Mk2A, although only eight are operational with XV704 having been de-converted at a later date and replaced by Sea King AEW Mk2A (XV707).

The main feature of the Sea King AEW Mk2A is its Thorn EMI Search Water maritime surveillance radar which is housed in a rigid, air-pressurised 'kettledrum' container attached to a swivel mount on the starboard side of the aircraft. The radar container can be hydraulically swivelled up and down. When the radar is not in use it can be retracted allowing the necessary ground clearance as the radar housing reaches below wheel level when deployed. The radar scanner is both pitch and roll stabilised and offers full 360 degree scan. To cope with the additional power needed for the radar a third electrical alternator is fitted as well as the Racal MIR-2 'Orange Crop' Electronic Surveillance Measure (ESM) system which is standard on the HAS Mk5/6.

Manned by a single pilot and two observers the helicopter is capable of locating and classifying small fast moving targets against heavy sea and weather clutter. Targets can be detected at ranges well in excess of 100nm (185km) from an aircraft on station at 10,000ft (3,000m) and a Multiple Track-while-Scan capability permits continuous tracking without interrupting search. A loitering speed of around 90kts with the radar deployed allows the aircraft an endurance of over four hours which can be extended further by helicopter inflight refuelling (HIFR) off nearby ships. 'Orange Crop' ESM, secure speech UHF/HF/VHF radios and data links enables fast warning of a target's course, speed, coded identity, range and bearing to be transmitted. The Sea King AEW Mk2A gives the fleet long range, over-the-horizon warning of threats posed from low flying missiles and aircraft. Under the control of the AEW Observer they can also deploy Sea Harriers to find and attack these targets.

849 Naval Air Squadron has its Headquarters and Training Flight based at RNAS Culdrose with two flights deployed aboard Royal Navy aircraft carriers. 'A' Flight is embarked in HMS *Invincible* and 'B' Flight aboard HMS *Ark Royal*.

All Airborne Early Warning helicopter pilots must have previously completed at least two tours of duty in other Sea King Squadrons. As no aircrewmen work in this specialisation, the squadron's AEW courses only train observers. Operating the Sea King AEW Mk2 as single pilot requires one of the observers to occupy the front left-hand seat (Co-Pilot) during landings, take-offs and during low level operations. One of their tasks is to operate manual throttles should they encounter a computer freeze, runaway or instability, or any other inflight emergency.

Opposite: Two 849 Squadron Sea King AEW Mk2As flying in formation over Somerset in July 1991. Sea King XV697 was converted to the AEW role on 23 November 1984. *(Patrick Allen)*

Below: Sea King AEW Mk2A (XV671) first flew on 26 August 1970 as a HAS Mk1 and first flew as an AEW Mk2A on 31 January 1985, joining 849 Naval Air Squadron on 5 March 1985. *(Westland)*

187
87
ROYAL NAVY
DANGER

The AEW observer training course includes six weeks spent at the Fighter Control School at RNAS Yeovilton and the complete course takes around five months to complete. Much of this training is carried out in flight over the South Western Approaches and the English Channel. Hunter and Falcon aircraft operating out of Yeovilton and Bournemouth co-operate in these vital training exercises.

849 Squadron headquarters saw the introduction of an Airborne Early Warning Simulator in 1989 which has helped to reduce the amount of flying time during training and has become a valuable aid to all forms of tactical and operational training. The Royal Navy AEW observer course is one of the most expensive courses in the Fleet Air Arm. Graduates from these courses immediately join the operational flights of 849 Squadron and deploy in aircraft carriers to provide the 'Eyes for the Fleet'.

SEA KING AEW Mk2A

Crew: 1 Pilot, 2 AEW Observers.
Power Plant: 2 Rolls-Royce Gnome H1400-1, 1,660shp.
Weight: Max AUW 21,400lb.
Performance: Max speed 112kts.
Cruising Speed Radar Deployed 90kts.
Endurance 4 hours plus.
Role Equipment: Thorn-EMI Search Water Radar inc. Third Alternator (weight approx. 3542lb). HF/VHF/UHF Radios, IFF/Data Links. Racal MIR-2 ESM.

SEA KING AEW Mk2A

Nine aircraft converted to AEW Mk2A standard with the first conversion taking place on 23 July 1982 (XV704) followed by: XV649, XV650, XV656, XV671, XV672, XV697, XV707 and XV714.

Sea King XV649 first flew on 21/9/69 and served with the original 700S IFTU before joining 706 Squadron at RNAS Culdrose. It was converted to an AEW Mk2A and joined 849 Naval Air Squadron in 1986.

Below: An AEW Mk 2A Sea King from 849 Naval Air Squadron with HMS *Invincible* in the background, launching a Sea Harrier FRS1. Since the Falklands Conflict the AEW Mk 2 Sea Kings have been the 'Eyes of the Fleet'. *(Royal Navy)*

SECTION 14

ROYAL AIR FORCE SEA KING HAR Mk3

Having re-equipped with the Wessex HC 2 in the mid-1970s (22 Squadron), the Royal Air Force wanted to replace their fleet of SAR Westland Whirlwind HAR 10s with a long-range SAR helicopter. The Westland Sea King had already built a reputation as a capable long-range maritime Search and Rescue helicopter and on 24 September 1975 the RAF ordered fifteen Sea King HAR Mk3s. A further four aircraft were built, one in 1980 and three in 1985 bringing the total to 19 aircraft.

The RAF Sea King HAR Mk3 retained the larger rear cabin of the West German Navy Mk41 and Norwegian Air Force Mk43s allowing the carriage of 17/19 passengers or six stretchers in the casevac role plus a crew of four. Powered by two Rolls-Royce Gnome H1400-1 turboshaft engines developing 1,535shp the HAR Mk3 had an all-up weight of 21,000lb. Larger fuel tanks allowed the aircraft to carry a maximum of 6,371lb of fuel giving a radius of action of approximately 280nm allowing 30 minutes on scene and 30 minutes reserve. The HAR Mk3s were also fitted with an improved cabin door-mounted rescue winch.

A full IFR-equipped cockpit, an Automatic Flight Control System (AFCS) and ARI5955 MEL lightweight radar allowed the Sea King HAR Mk3 to undertake day or night IMC operations over the sea. The comprehensive navigation systems which have been regularly updated include: VOR, ADF, DME, ILS, VHF and UHF radios, Decca Mk19 and Omega navigation computers which can be fed into the Doppler TANS. The AR1 5955 Search Radar has a radius of approximately 50 miles and can be used to navigate,

Below: Lossiemouth-based 202 Squadron Sea King HAR 3 (XZ590) was one of several RAF SAR Sea Kings to assist during the Piper Alpha disaster. *(Patrick Allen)*

locate survivors/targets or undertake search patterns. The aircraft are also fitted with UHF and VHF homers (500 kHz and 2182 kHz) and some are also fitted with Chelton VHF/FM maritime channel 16 and channel 67 homers. Six of the RAF Sea Kings are painted grey and are used by 78 Squadron in the Falkland Islands. The squadron has two aircraft based at RAF Mount Pleasant with the remainder in transit/maintenance or operating with the Sea King Training Unit (SKTU) at RNAS Culdrose.

In addition to the yellow aircraft of 202 Squadron, the grey Sea Kings of 78 Squadron are fitted with Radar Warning Receivers, Omega Navigation systems, Chelton Homers and Night Vision compatible cockpits. Other systems due to be fitted or have recently been fitted included an Automatic Voice Alerting Devise (AVAD). This is a computer activated voice, used to alert pilots giving them flight safety voice warning and advisory messages, through the use of pre-recorded, digitally stored human speech activated in potentially dangerous situations. Messages can warn a crew by telling them to check their height, 'Check height', 'Check Landing Gear', 'Master Caution', 'Fire Warning Port/Starboard'. AVAD was fitted to RAF Puma and Chinook helicopters during the Gulf War and was found to be particularly useful in high work-load situations particularly when using Night Vision Goggles and operating at low heights. Other modifications include the fitting of the Global Positioning System (GPS) Navastar Racal RNS 252 (Supertans) which integrates both Doppler and GPS navigation systems.

The Sea King HAR Mk3 (XZ585) first flew on 6 September 1977 and joined No 202 Squadron replacing the Whirlwind HAR Mk10. With its Headquarters and SAR Engineering Wing based at RAF Finningley, 202 Squadron detached two Sea King HAR Mk3 Flights initially to Boulmer, Northumberland, Brawdy, Coltishall and Lossiemouth, Scotland. More recent changes have included Flights being based at:

SAR Wing HQ, RAF Finningley.
(RAF Boulmer 1993) 202 Squadron:
- 'A' Flight, RAF Boulmer, Northumberland.
- 'B' Flight, RAF Chivenor, Devon.
- 'C' Flight, RAF Wattisham, Suffolk.
- 'D' Flight, RAF Lossiemouth, Grampian.
- 'E' Flight, RAF Leconfield, Humberside.

78 Squadron: 2 Sea King HAR Mk3s at RAF Mount Pleasant, Falkland Islands.
Sea King Training Unit (SKTU): Moved to RAF St Mawgan 1993. 2 Sea King HAR Mk3s, RAF St Mawgan, Cornwall.

All RAF SAR helicopters including the Sea Kings are primarily tasked with military SAR but the vast majority of their missions involve civilian SAR tasking both on land and at sea. In the UK, SAR operations are overseen by two Rescue Co-ordination Centres (RCCs) one at Pitreavie (Edinburgh) and the other at Plymouth in Devon. The two centres receive and collate all information on incidents, initiate helicopter call-outs and update the relevant services involved. These can include RAF Nimrod aircraft, Mountain Rescue Teams, Police, Coastguard and the RNLI etc. RAF Sea King Flights have one helicopter and crew held at 15 minutes notice during daylight hours and 45 minutes at night with a second crew available at short notice. All the Sea King Flights are kept extremely busy undertaking day and night rescue missions which includes operating in the mountains of Scotland and Wales and at sea. Over the years, the Sea Kings of 202 Squadron have undertaken some of the longest recorded maritime rescues many far out into the Atlantic.

RAF SEA KING TRAINING UNIT

Based in Cornwall, the RAF Sea King Training Unit (SKTU) is responsible for the conversion and training of all RAF Sea King crews. Originally formed in 1978 the Unit undertook the initial training of the first 202 Squadron crews and today, as part of the Rescue Wing of the RAF's No 18 Group, they train all RAF Sea King crews.

The SKTU has two Sea King HAR Mk3s

Below: 78 Squadron Sea Kings are fitted with a Chelton Homer and Omega navigation. Future updates to the RAF Sea King fleet may include AVAD, Maritime Homers and Satellite GPS navigation system. The six new build RAF Sea King HAR 3s will have an improved flight path control and 'auto hover' AFCS plus a digital colour search radar and possibly a FLIR. *(Patrick Allen)*

and a staff of four Qualified Helicopter Instructors (QHIs), one of which is the Commanding Officer, three Qualified Crewman Instructors (QCIs) plus Groundcrew and Administrative Staff of approximately thirty. The Unit trains approximately six full crews per year (Pilots, Winch/Radar Operator and Winchman; both rear crews are cross trained) and also undertake Pilot refresher training (one month), Senior Officers familiarisation (one week), OC 78 Squadron familiarisation (one week. As a two-aircraft Squadron operating Chinooks and Sea King the OC 78 Squadron may only have Chinook experience), Rear Crew refresher training (two weeks), QHI/QCI Sea King conversion (one month).

The 90-hour Sea King course is broken down into five stages with both front and rear crews training together. The five stages involve Aircraft Handling, Instrument Flying, Night Flying and AFCS auto-transitions, Role Conversion which includes winching to boats, cliff winching, operating at night, locating and recovering survivors in single and multi-seat dinghies day and night and the use of the High-Line winching method. The advance phase and final exercise titled 'Yellow Scorpion' includes Deck Landing Practise (DLPs), underslung load lifting and final three-day exercise. Students are put into the position of being on Search & Rescue standby and then complete several SAR scenarios. Both student pilots take turn as Captain and the rear crew carry out one day as winch/radar operator and the other as winchman. Students are sent on a number of different types of rescue missions which includes mountain flying training in Snowdonia. On completion of the course, crews are sent to one of 202 Squadron's SAR Flights where they carry out their Squadron acceptance before becoming part of the Duty SAR crew. Pilots have to be fully experienced before becoming aircraft captains. This usually takes 18 months on an Operational Flight. Both students and operational crews make full use of the Sea King Simulator based at RNAS Culdrose.

Below: Winching training by a SKTU Sea King HAR 3. Both front and rear crews train together making maximum use of valuable aircraft hours. RAF aircrews will within the next few years benefit from their own Sea King flight/mission simulator to be based at RAF St Mawgan. *(Patrick Allen)*

78 SQUADRON

Based at RAF Mount Pleasant in the Falkland Islands 78 Squadron operates two Sea King HAR Mk3s which provide search and rescue cover for all military operations in and around the Falkland Islands, as well as aiding in civilian rescues and other emergency tasks. The Squadron also operates the Chinook helicopter and both aircraft are able to support each other in all the Squadron's roles.

During the Falklands War a single Sea King HAR Mk3 (XZ593) from 202 Squadron was airfreighted on 8 May 1982 to Ascension Island where it undertook HDS and SAR missions. The Sea King frequently flew over 12 hours a day and undertook several casevac missions including one lasting over eight hours. The Sea King HAR Mk3 stayed at Ascension until September when it was shipped back to the UK.

After the War, the RAF decided to deploy three Sea King HAR Mk3s to the Falkland Islands to provide SAR cover and general logistics support. Three Sea Kings (XZ591, XZ592 and ZA105) plus personnel from 'C' Flight 202 Squadron were dispatched to the Falklands arriving in theatre on 25 August 1982. All three Sea Kings were modified for the Falkland Islands operations. They lost their familiar SAR Yellow and were resprayed in Dark Sea Grey. Enhancements included Night Vision Goggle (NVG) compatible cockpit lighting, Omega navigation equipment and Radar Warning Receivers (RWR) plus additional military radios.

'C' Flight 202 Squadron operated the 'SARDET' from Navy Point from 25 August 1982 until the Flight was re-designated 1564 Flight on 20 August 1983. On 22 May 1986 the three Sea Kings which had been regularly rotated back to the UK for major maintenance, joined the Chinooks from 1310 Flight to form 78 Squadron. Today the Squadron's two Chinook HC Mk1s and two Sea King HAR Mk3s operate from RAF Mount Pleasant.

The two Sea Kings provide day and night SAR cover with one aircraft held at 15 minutes readiness by day and one hour

at night. They frequently undertake both military and civilian SAR missions as well as undertaking support helicopter duties including moving underslung loads of stores/supplies or carrying up to 12 fully equipped troops.

The Squadron's Sea Kings frequently undertake long range over the water SAR missions assisting fishing boats and merchant ships in distress. On two occasions they have undertaken rescue missions to South Georgia (800 nm). Using a ship to land on and refuel, the Sea King can be carried for a distance before leap-frogging onto South Georgia. The isolated position of the Falklands has provided the Sea King crews with ample experience of operating far into the South Atlantic with only the C130 Hercules of 1312 Flight to provide top cover. To undertake these long range missions the Sea Kings need to refuel from ships operating in the area and the crews regularly deck land or HIFR from local vessels which include RFA ships, Royal Navy Guard ships or Fishery Protection Vessels and Patrol Boats.

The versatility and airmanship of the 78 Squadron Sea King crews are fully exploited in the South Atlantic. On 26 November 1991 the Squadron received the 'Silk Cut Nautical Award' for their part in the rescue of 21 crew from the RFA *Gold Rover*. RFA *Gold Rover* lost her rudder and steerage in high winds and in sea state 9 conditions. The extreme weather conditions and rough seas resulted in the helicopter having to make several transits to the ship. The Sea King eventually had to High-Line transfer 11 crew aboard the helicopter as conditions deteriorated.

Another typical 78 Squadron rescue mission occurred in May 1990 when a Sea King was launched on a 320 nm transit into the South Atlantic to recover an injured crewman from a fishing boat. Weather conditions were poor and the outside air temperature was minus 2 degrees. Refuelling aboard HMS *Leeds Castle*, and with top cover provided by a 1312 Flight C130K, the fisherman was safely delivered to Port Stanley Hospital. The mission lasted 7 hr 20 mins which included one hour night flying. As well as undertaking maritime rescues the Sea Kings are also kept busy with land-based

Top left: A 78 Squadron SAR Sea King (XZ599) operating in typical bleak Falkland Island countryside. All 78 Squadron SAR Sea Kings are painted Sea Grey for operational reasons and are fitted with NVG cockpit lighting, Omega navigational equipment and RWR. *(Patrick Allen)*

Bottom left: Falkland-based 78 Squadron Sea Kings often deck land and HIFR from ships operating in the South Atlantic. Here a 78 Squadron Sea King exercises with the Port Stanley based MV *Bransfield* during a winching exercise. *(Patrick Allen)*

rescues, locating lost patrols, injured soldiers and civilians, and undertaking casevac missions throughout the islands.

The island's mountains, plus strong winds and changeable weather, requires expert mountain flying techniques and crews must be capable of operating in marginal weather conditions. The RAF Sea Kings can accommodate 17/19 survivors and land-based night searches are further enhanced by crews wearing Night Vision Goggles (NVGs). NVGs are useful for overland transits, particularly during mountain flying operations or in poor weather. Another role for the Falkland Sea King is as firefighters using the Simms Rainmaker fire bucket which can lift one ton of water. During September 1991 a Sea King helped to put out a major heath fire which was threatening a Rapier Missile site located close to the airfield.

On Wednesday 19 February 1992 the Minister of State for Defence Procurement announced in the House of Commons an order for six new Royal Air Force Sea King HAR Mk3 helicopters for the SAR role to replace the ageing UK-based Wessex in the search and rescue role. The order also included a Sea King HAR Mk3 flight simulator which will now enable both front and rear RAF Sea King crews to undertake full mission simulator training adapted for the SAR role in their own dedicated simulator instead of having to use and adapt the Royal Navy's Sea King HAS Mk5 simulator based at RNAS Culdrose in Cornwall.

The new build RAF Sea Kings will probably be designated as Sea King HAR Mk3As and new equipment may include an uprated digital colour Search Radar, an improved Automatic Flight Control System (AFCS) to include a fully-coupled flight path control and 'auto-hover' system, a nose-mounted Bendix weather radar, forward-looking infrared (FLIR) plus the Racal RNS 252 GPS Supertans navigation system.

SEA KING HAR Mk3
19 a/c built

XZ585 (6/9/77) to XZ599 (1/12/78) 15 a/c
ZA105 (14/ 8/80) 1 a/c
ZE368 (22/5/85) to ZE370 (21/ 8/85) 3 a/c

Top right: 78 Squadron SAR Sea King HAR 3 (XZ591) in front of an Argentine 155mm howitzer at its home base RAF Mount Pleasant. *(Patrick Allen)*

Bottom right: An RAF SAR Sea King HAR 3 operating with the RAF Sea King Training Unit (SKTU) based at RNAS Culdrose. This unit moved to RAF St Mawgan in December 1992 as RAF SAR Squadrons redeploy during 1993/94. *(Patrick Allen)*

SECTION 15

ROYAL NAVY SEA KING HAS Mk5

A continual Sea King updating programme by Westland plus the rapid advance in the development and integration of electronic mission equipment led to the Sea King HAS Mk5. These advances resulted in the Royal Navy ordering 30 new build Sea King HAS Mk5s and the conversion of the majority of their existing fleet of HAS Mk2s. The conversion programme was undertaken by the Royal Navy at both RNAY Fleetlands and RNAS Culdrose.

First flight of a Sea King HAS Mk5 (XZ916 a converted HAS Mk2) took place on 1 August 1980 with the first new build Sea King HAS Mk5 (ZA126) flying on 26 August 1980. Modifications included a larger rear cabin area, strengthened cabin floor, larger more powerful digital MEL Sea Searcher radar and improved navigational equipment to include a Decca TANS 9447G linked to a Decca Doppler 71. Mission equipment improvements included the installation of the Racal MIR-2 'Orange Crop' ESM equipment, passive sonobuoy dropping equipment and GEC Avionics LAPADS lightweight acoustic processing and display equipment. A number of Sea King HAS Mk5s were additionally fitted with MAD equipment which was housed in the starboard wheel sponson.

The new more powerful MEL Sea Searcher radar gave the HAS Mk5 almost double the search range of the previous Mk2 with its Ecko AW 391 radar. Sonobuoys allowed the Sea King to spread its ASW net far wider than ever before. It could not only monitor its own sonobuoys, but could also monitor and collate information from buoys dropped by RAF Nimrod aircraft. Capable of operating for long periods at a radius of over 100 miles from its parent ship, the Sea King HAS Mk5 could now pinpoint and attack submarines with up to four Mk46/Sting Ray advanced lighweight torpedoes or Mk11 depth charges at far greater distances than had been possible before.

'OPERATION CORPORATE'

The first Sea King HAS Mk5s entered service with No 706 Naval Air Squadron based at RNAS Culdrose and with No 820 Squadron in June 1981 embarking in HMS *Invincible*. Within a year the Sea King HAS Mk5 was in action protecting the British Task Force during the Falklands War.

No 820 Naval Air Squadron embarked in HMS *Invincible* and No 826 Naval Air Squadron embarked in HMS *Hermes*. They provided a continuous ASW screen for the Task Force which involved three Sea Kings flying ahead of the Carrier Group providing an ASW screen with a fourth Sea King further out undertaking surface search screening out to distances of over 200 miles.

Below: 820 Naval Air Squadron Sea King HAS Mk5s line up on HMS *Invincible*. *(Westland)*

During one record-breaking sortie one Sea King was airborne for 10hr 20mins, crew changing and refuelling in the hover (HIFR). Once the Task Force was on station in the Falklands, the Sea King HAS Mk5s continued their ASW screening and helped to enforce the 200-mile Total Exclusion Zone.

The Sea King HAS Mk5 proved outstanding during the conflict and aircraft availability exceeded 85 per cent. 820 Naval Air Squadron flew 2,150 hours on 700 ASW sorties as well as undertaking other missions which included SAR. 826 Naval Air Squadron totalled 3,000 flying hours and both Squadron Commanders received the Air Force Cross in recognition of their Squadrons' outstanding contribution.

The Sea King HAS Mk5 continued to live up to its reputation as an outstanding maritime ASW and SAR helicopter. In 1987 they helped transport divers and equipment to Belgium and assist at the *Herald of Free Enterprise* disaster. In 1988 the Royal Navy replaced their SAR Wessex helicopters with a modified Sea King HAS Mk5. With the ASW equipment removed, but retaining the MEL Sea Searcher radar, the Sea King Mk5 provided the Royal Navy with an all-weather day and night SAR helicopter.

During 1989, 826 Naval Air Squadron who now provided Sea King helicopters for deployment onboard Royal Fleet Auxiliaries began trials to discover whether their Sea King Mk5 aircraft could operate and fly successfully from the Navy's Type 22 Frigates. Three months of intensive flying trials proved successful and the Sea King HAS Mk5s from 826 Naval Air Squadron are now permanently assigned to Royal Navy Type 22 Frigates.

OPERATION GRANBY/OPERATION MANA

The Sea King HAS Mk5 saw action once again during 'Operation Granby' when two modified 826 Squadron 'D' Flight Sea King HAS Mk5s deployed to the Gulf in RFA *Olna* in August 1990. Operating as part of the multi-national team enforcing sanctions against Iraq, their tasks included locating, searching and questioning merchant shipping traffic operating in the Gulf. The 826 Squadron Sea King HAS Mk5s had their ASW equipment removed and were modified to undertake the surface search role and they later proved vital in the mine spotting role.

Top right: Exercise 'Purple Warrior' 1987 and Sea King HAS Mk 5 (XZ582) lands on HMS *Illustrious* behind a Sea King HC Mk 4. XZ582 was written-off on 27 October 1989 when it ditched off Bermuda and sank. *(Patrick Allen)*

Right: Royal Navy Sea King HAS Mk 5s and 6s line up at RNAS Yeovilton during the Presentation of the Colour to the Naval Air Command at HMS *Heron* by Her Majesty The Queen on 27 June 1991. *(Patrick Allen)*

Modifications to the aircraft included the fitting of a video camera system, Sandpiper Thermal imaging system (FLIR), NAVSTAR Global Positioning System (GPS), secure speech radios and a comprehensive Defence Aids Suite.

'D' Flight was relieved by 'C' Flight in December and deployed aboard the Dutch RFA, HNLMS *Zuiderkruis*. Shortly after the Gulf War began in mid-January 'C' Flight deployed to RFA's *Sir Galahad* and RFA *Argus* operating in the anti-mine role. By February they were operating in the Northern Gulf and successfully located a number of mines which were destroyed by despatching divers. By 27 February 'C' Flight were supporting Special Forces troops entering Kuwait City Marina. After the ceasefire on 28 February 1991 'D' Flight continued their hunt for mines and moved to RFA *Fort Grange* in mid-March to be replaced by 'C' Flight in late April. Within a few weeks 826 Squadron Sea Kings aboard RFA *Fort Grange* were dispatched to Bangladesh to provide humanitarian assistance to those in need. 'Operation Mana' was completed by 3 June 1991 and the four Sea Kings (2/ 826 Squadron 2/845/846 Squadron) moved over 400 tons of relief supplies.

Other Royal Navy Sea King squadrons based at RNAS Culdrose were equally as busy during the Gulf War. On 10 January 1991 820 Naval Air Squadon embarked on HMS *Ark Royal* with 10 Sea Kings and sailed for the Eastern Mediterranean. Establishing a patrol area off Cyprus, the squadron undertook the surface search role and vertical replenishment of the Task Group which included HM Ships *Ark Royal*, *Manchester*, *Sheffield*, RFAs *Olmeda* and *Regent*. One of the squadron's more important tasks was to provide long range surveillance in the vicinity of the entrance to the Suez Canal which was vital for the reinforcement of allied forces who were operating in the Red Sea, Indian Ocean and Persian Gulf.

Top right: Sea King aboard HMS *Ark Royal*. *(RNAS Culdrose)*

Right: 826 Squadron Sea King HAS Mk 5 from HMS *Coventry's* flight. The new Merlin EH101 will replace the Sea King on ships' flights. *(RNAS Culdrose)*

SEA KING HAS Mk5

Crew: 4 (2 Pilots, 1 Observer, 1 Sonar Operator).
Power Plant: 2 Rolls-Royce Gnome H1400-1, gas turbine engines developing 1,600shp (take-off) and max. continuous 1,250shp.
Dimensions: Rotor diameter 62ft (18.9m). Length 55.8ft (17.01m).
Performance: HIGE 5,000ft (1,525m). HOGE: 3,200ft (975m). VNE: 144kts (267km/h). Max cruise 112kts (208km/h). Range 664nm (1,230km) full standard fuel. Max AUW 21,400lb (9,707kg).
Weapons: 4 × 44/46/Sting Ray torpedoes, depth charges, door guns.

WESTLAND SEA KING HAS Mk5 — 30 built

ZA126 (26/ 8/80) to ZA137 (17/ 7/81) — 12 a/c
ZA166 (4/ 3/82) to ZA170 (2/ 9/82) — 5 a/c
ZD630 (9/ 8/84) to ZD637 (23/ 4/84) — 8 a/c
ZE418 (22/ 1/86) to ZE422 (10/ 7/86) — 5 a/c

SEA KING HAS Mk5 SQUADRONS

706 NAVAL AIR SQUADRON,
RNAS CULDROSE
(Sea King ASW crew conversion)

810 NAVAL AIR SQUADRON,
RNAS CULDROSE
(Advanced flying and
ASW operational training)

814 NAVAL AIR SQUADRON,
RNAS CULDROSE
(HMS *Illustrious* and HMS *Invincible*)

819 NAVAL AIR SQUADRON,
HMS *GANNET*, PRESTWICK
(ASW/Defence of Clyde SSBN base/SAR)

820 NAVAL AIR SQUADRON,
RNAS CULDROSE
(HMS *Invincible* and HMS *Ark Royal*)

824 NAVAL AIR SQUADRON,
RNAS CULDROSE
(RFA Sea King Flights)

826 NAVAL AIR SQUADRON,
RNAS CULDROSE
(HMS *Hermes*, RFA and Type 22 Frigates)

Top left: Sea King HAS Mk 5 — 826 NAS — *Coventry* flight. *(RNAS Culdrose)*

Left: A Sea King HAS Mk 5 from 814 Squadron flies with a 'MAD' sponson. HMS *Invincible* is in the background. *(RNAS Culdrose)*

Top right: A crewman about to drop a sonobuoy from a Sea King HAS Mk 5. *(Westland)*

Bottom right: The rear cabin and sonar equipment of an ASW Sea King HAS Mk 5. *(Westland)*

Below: An early Sea King HAS Mk 5 from 820 Squadron embarked on HMS *Invincible* on a dunking sortie in the hunter/killer role carrying Mk 44 torpedoes. *(Westland)*

SECTION 16

ROYAL NAVY SEARCH & RESCUE SEA KINGS

Royal Navy Search and Rescue became an all-Sea King operation in 1988 when both 771 Naval Air Squadron based at RNAS Culdrose re-equipped with the Sea King Mk5 and 772 Naval Air Squadron at RNAS Portland re-equipped with the Sea King HC Mk4. Both squadrons were the last Royal Navy squadrons to operate the Wessex HU Mk5 and this change marked the retirement of the venerable Wessex from the Fleet Air Arm. 819 Naval Air Squadron based at HMS *Gannet*, Prestwick, had been operating a Sea King SAR Flight since November 1971. Their two ASW Sea King Mk5/6s had been providing SAR cover extending from the north of England to the north of Scotland.

SEA KING SAR

From their introduction into the Fleet Air Arm, the Sea King began establishing an unbeatable reputation as an outstanding SAR helicopter. The specifications for a successful ASW helicopter, i.e. long endurance and the ability to operate independently over the water in all weathers, day and night were also the main ingredients for an outstanding SAR helicopter. A spacious cabin which could accommodate large numbers of survivors (23) plus powerful Rolls-Royce gas turbine engines giving long range and reliability, further enhance the aircraft's versatility.

Although Royal Navy SAR squadrons and flights were equipped with the Wessex there was always the ability to call upon either shore or ship-based RN Sea Kings to assist if required. The success of the Sea King and its increased reputation as a SAR helicopter led, in 1975, to the Royal Navy making their shore-based Sea Kings at RNAS Culdrose and Prestwick permanently available for SAR taskings, as well as undertaking their normal operational duties. Almost every year since their introduction into service, Royal Navy Sea King helicopters have been involved in headline rescues both at home and abroad. The bravery and skill of their crews is evident in the number of Gallantry Medals and citations awarded. These included a number gained during the Falklands Campaign where Sea Kings undertook numerous rescue missions including the rescue of survivors from the RFA *Sir Galahad*.

819 NAVAL AIR SQUADRON HMS *GANNET*, PRESTWICK

819 Naval Air Squadron is an ASW Sea King squadron. They are tasked with providing ASW cover and support to the Clyde-based SSBN nuclear submarines and for providing a Search and Rescue Flight. The squadron are equipped with the ASW Sea King HAS Mk5/6. The two helicopters assigned to the SAR Flight have some of their sonar equipment removed which helps to reduce aircraft weight and extend endurance. Providing a 24-hour SAR cover, the squadron area covers a distance from the north of England up to the north of Scotland and extends west to around 200nm into the Atlantic north of Ireland. Providing both military and civilian SAR cover, the majority of callouts (90 per cent) are civilian emergencies. The squadron is tasked by the Maritime Rescue Co-ordination Centre at RAF Pitreavie and maintains an aircraft on 15 minutes' standby in daylight hours and 45 minutes at night.

To overcome the weight penalty of carrying ASW/sonar equipment, although some of this is removed, the squadron has established their own fuel dumps throughout Scotland. This enables their aircraft to operate for extended periods away from base, and increases their operational range. This also allows the squadron to react quickly to SAR call-outs, even when undertaking other operational duties. Other options for obtaining fuel in the area include refuelling from RN or RFA ships/oil rigs etc. During 1988 the squadron had 76 callouts, this increased to 83 in 1989 and by 1990 there were over 142 scrambles.

Below: 771 Squadron SAR Sea King Mk 5 (XZ920) working with the RNLI lifeboat. Based at RNAS Culdrose, Sea King Mk 5 (XZ920) first flew in 1979 as a HAS Mk 2. *(RNAS Culdrose)*

771 NAVAL AIR SQUADRON SEARCH & RESCUE

771 Naval Air Squadron became a dedicated shore-based SAR squadron at RNAS Portland in 1969 operating the Wessex Mk1. On 4 September 1974 the squadron moved to RNAS Culdrose and in March 1979 they re-equipped with the Wessex HU Mk5. The squadron was responsible for the SAR commitment at RNAS Culdrose and for training SAR aircrew. This included running the Royal Navy SAR Diver and Winchman courses. The SAR Diver is an important factor in the success of the Royal Navy SAR helicopter and is unique to them. Royal Navy SAR divers attend a four-week Ship's Divers Course covering day and night training at HMS *Vernon*, Portsmouth, before going to RNAS Culdrose for a five-week SAR divers' course. These divers, who work unattached from the helicopter are capable of jumping from heights in excess of 40 feet above the sea and can dive down to get crews out of submerged aircraft, helicopters etc. The flexibility and manoeuvrability of a diver, even in the most adverse sea conditions and their ability to move freely through or under the water during a rescue, has resulted in the saving of many lives. SAR divers are also trained aircrewmen and can navigate the aircraft as well as operate and descend on the winch.

Below: 771 Squadron Sea King Mk 5 (XV705/821) seen in early May 1988 having just joined the squadron. The aircraft is seen wearing the squadron's 'Club' card motif but without its red painted nose and tail. *(Patrick Allen)*

In the spring of 1988, the squadron re-equipped with the Sea King Mk5. This was basically a Sea King HAS Mk5 but with the ASW/Sonar equipment removed greatly reducing the aircraft's weight. This resulted in an extremely capable, long range SAR helicopter which could accommodate up to 23 survivors. The advanced Automatic Flight Control System (AFCS) providing automatic transition and hover facilities, enables the helicopter to operate over the water, at night, in extremely adverse conditions. It also allows the winchman working from the cabin door a ten per cent over-ride control during an AFCS hover. All this, together with the HAS Mk5's long range MEL Sea Searcher radar allowing efficient search patterns to be followed, gives the Sea King Mk5 a true all-weather day and night capability.

One item which has gone a long way to make the Sea King such a successful SAR helicopter is the door-mounted winch or hoist. The hydraulically-operated hoist can be controlled either by the pilot or the crewman, depending on the position of a hoist master switch located on the cockpit roof panel. The hoist has a variable speed control allowing winch speeds of up to 200ft/m. The 245ft long cable has a maximum weight limit of 600lb. In an emergency the cable can be cut (Hoist Cable Cutter) by either the pilot or winch operator.

The Sea King Mk5 enables 771 Naval Air Squadron to provide 24-hour, day and night SAR cover for both the military and civilian authorities. The squadron is required to be able to scramble an aircraft within 15 minutes by day and 45 minutes by night. In reality these times are usually reduced to four minutes during the day and 15 minutes at night. Providing 24-hour cover 365 days a year the squadron SAR area extends to over 200nm.

In their first year of Sea King operations the squadron undertook a record 155 missions responding to over 167 callouts. In 1989 the squadron had another record-breaking year with over 192 missions when 216 people were either assisted or saved. During 1989, the squadron were involved in several headline rescues which included the rescue of the MV *Secil Japan* and the rescue of MV *Murree*. Three 771 Squadron aircrew-man/divers were awarded the Air Force Medal and the George Medal for their part in these rescues.

772 NAVAL AIR SQUADRON
SEA KING HC Mk4 (SAR)

Based at RNAS Portland, 772 Naval Air Squadron was reformed out of the Wessex Mk1s of 771 Naval Air Squadron when they moved down to RNAS Culdrose in September 1974. The Wessex Mk1s were replaced by the Wessex HU Mk5 in September 1977.

After the Falklands War in 1982, the squadron became responsible for providing the SAR cover at HMS *Daedalus*, Lee-on-Solent and this they did with a detached flight of two Wessex. With the end of the Wessex in March 1988, the Lee-on-Solent Flight was disbanded and the Portland Military SAR cover was undertaken by the squadron's new Sea King HC Mk4s based at RNAS Portland. The first of five new Sea King HC Mk4s arrived at RNAS Portland in late February 1988 and took over from the Wessex in both the SAR and personnel movements (SOOTAX) service for the Flag Officer Sea Training (FOST) which is undertaken by the squadron.

771 Squadron's SAR responsibility covers both military and civil call-outs and extends to an area from Brighton to Plymouth on the South Coast and inland to Oxford. 772 Squadron's Sea Kings differ from other Commando squadron helicopters. All Sea King HC Mk4s are fitted with an Automatic Flight Control System (AFCS). In the helicopter's Commando role these are 'gagged' out. To allow the SAR Sea King HC Mk4 to hover over the water at night, the AFCS has been reinstated. This allows the pilot to engage the AFCS at a given height and to automatically transition the helicopter down to a chosen hover height. This facility also gives the winchman the ten per cent over-ride control available on other Sea Kings. Other additions to the SAR Sea King HC Mk4 is the fitting of a plastic sea tray for the rear cabin, a Pye Beaver 12 channel maritime radio and a Chelton VHF Maritime Homer. This allows the helicopter to home-in immediately on a VHF transmission. The Commando Sea King HC Mk4 can carry 23 seated passengers at a speed of over 100kts and has an endurance in excess of four hours.

Operated single pilot by day and with two pilots at night — plus a crewman and SAR diver, the squadron undertook over 107 SAR callouts in their first year of Sea King operations. In 1989 they undertook over 121 callouts.

772 NAVAL AIR SQUADRON
New build Sea King HC Mk4s — 5 a/c

ZF120 (11/11/86)
ZF121 (2/12/86)
ZF122 (15/ 1/87)
ZF123 (25/ 3/87)
ZF124 (4/ 4/87)

Below: 771 Squadron Sea King Mk 5 (XV705/821) on a training mission in early 1988 along the Cornish coast. Having only just received the Sea King the squadron had not had time to paint the aircraft with the now familiar red nose and tail. The new Sea King Mk 5 gave the Royal Navy their first dedicated all-weather day and night SAR helicopter replacing the Wessex HU Mk 5 in this role. *(Patrick Allen)*

Right: 771 Squadron SAR Sea King Mk 5 (XZ920/822) first flew in May 1979 as a HAS Mk 2 and is seen here, blades folded, at RNAS Culdrose. *(Patrick Allen)*

Below: 771 Squadron SAR Sea King Mk 5 (XV647/820) was the fifth Sea King to be built at Westland and first flew on 6 September 1969. Originally joining the Sea King IFTU (1969/70) it now operates in the SAR role as a Mk 5 by 771 Naval Air Squadron based at RNAS Culdrose. *(Patrick Allen)*

Left: 772 Squadron Sea King HC Mk 4 (ZF124/24) seen during a winching practice in Portland Harbour. 772 Squadron Sea Kings operate single pilot with a crewman and SAR diver. During high work loads or night/NVG sorties a second pilot is used. *(Patrick Allen)*

Below: 21 March 1988 and the first of five brand new Sea King HC Mk 4s arrive at RNAS Portland to join 772 Naval Air Squadron to undertake SAR and FOST duties, replacing the Wessex. *(Patrick Allen)*

SECTION 17

SEA KING HAS Mk6

The Sea King HAS Mk6 is the latest of the Royal Navy's Sea King variants. It takes advantage of Westland Helicopters' continual updating programme of the Sea King to the latest Advanced Sea King standard. Similar in appearance to the Sea King HAS Mk5, the HAS Mk6 is fitted with Westland composite advanced technology rotor blades and an improved operational capability.

The Sea King HAS Mk6 is fitted with the advanced Plessey/GEC Digital Type 2069 Dipping Sonar which can operate down to the deepest levels in search of submarines. The Data from both the dipping sonar and sonobuoys is now integrated through a new digital GEC Avionics AQS-902G-DS processor. This includes both active and passive data and information from the MEL Sea Searcher radar, which all appear on a single CRT, reducing operator workload. The HAS Mk6 also has an improved radio fit which includes the GEC AD3400 secure U/VHF communications system. The HAS Mk6 has greatly improved the Royal Navy's ASW capabilities until the arrival of the EH-101 'MERLIN'.

The first conversion of a Sea King HAS Mk6 (XZ581) took place on 16 April 1987. The Royal Navy ordered five new build Sea King HAS Mk6s and began a programme to update their existing fleet of HAS Mk5s. The HAS Mk6 entered service on 1 February 1988.

SEA KING HAS Mk6 — 5 Built. Remainder converted.

ZG816 (7/12/89) to ZG819 (10/5/90) — 4 a/c
ZG875 (27/6/90) — 1 a/c

SEA KING HAS Mk6 SQUADRONS

810 NAVAL AIR SQUADRON,
RNAS CULDROSE
(Operational Flying Training ASW aircrew)

814 NAVAL AIR SQUADRON,
RNAS CULDROSE
(HMS *Invincible*)

819 NAVAL AIR SQUADRON,
HMS *GANNET*, PRESTWICK
(ASW/Defence of Clyde SSBN/SAR)

820 NAVAL AIR SQUADRON,
RNAS CULDROSE
(HMS *Ark Royal*

824 NAVAL AIR SQUADRON,
HMS *GANNET*, PRESTICK
(Sea King HAS Mk6 IFTU)

825 NAVAL AIR SQUADRON 'E' FLIGHT
Formed at A&AEE Boscombe Down
in the summer of 1991 with
two aircraft to evaluate new tactics
for the Sea King HAS Mk6.

SEA KING SQUADRONS OF THE NAVAL AIR COMMAND 1991/92

Sqdn	Aircraft Type	Shore Base
706	Sea King HAS Mk5/6	Culdrose
707	Sea King HC Mk4	Yeovilton
771	Sea King Mk5	Culdrose
772	Sea King HC Mk4	Portland
810	Sea King HAS Mk5/6	Culdrose
814	Sea King HAS Mk6	Culdrose
819	Sea King HAS Mk6	Prestwick
820	Sea King HAS Mk6	Culdrose
826	Sea King HAS Mk5/6	Culdrose
845	Sea King HC Mk4	Yeovilton
846	Sea King HC Mk4	Yeovilton
849	Sea King AEW Mk2A	Culdrose
848	Sea King HC Mk4	Yeovilton

(formed for Gulf War only)

824 NAVAL AIR SQUADRON SEA KING HAS Mk6 IFTU

824 Naval Air Squadron IFTU arrived at HMS *Gannet*, Prestwick, on 20 July 1987 with Sea King HAS 5s ZA134/252 and XV655/253.

The first Sea King HAS Mk6 ZA136/253 arrived at the IFTU on 15 April 1988 and the IFTU was disbanded in July/August 1989.

ROYAL NAVY ASW/AEW AIRCREW TRAINING

PILOT TRAINING

Having successfully completed aircrew selection at Royal Air Force Biggin Hill, students undertake 10 hours aptitude flying training in Chipmunks then a further 65 hours Elementary Flying Training on Bulldogs at RAF Topcliffe in Yorkshire. Students spend six months at RAF Topcliffe before moving to RNAS Culdrose to begin their Royal Navy helicopter training.

Here they join 705 Naval Air Squadron and undertake their Basic Flying Training (BFT) in the Gazelle HT2 helicopter. Students are brought up to 'Wings' standard after 22 weeks amassing a total of 81 flight hours, 25 per cent flown solo.

Those pilots streamed for the ASW Sea King join 706 Naval Air Squadron. Here they undertake a 16-week course and are allocated 54 Sea King flight hours. During this course

Below: A Royal Navy Sea King HAS Mk6 from 819 Squadron. *(RNAS Culdrose)*

students also fly 30 hours on the Sea King Mk5/6 simulator at RNAS Culdrose. The first Sea King Simulator arrived at Culdrose in 1971 and this was replaced in 1988 with an advanced six axis Mk5/6 Simulator. The new simulator has three rear modules for observers and aircrewmen which reproduces the radar and sonar stations of the HAS Mk5/6. By 1991 both pilot and rear crew modules had been updated to the HAS Mk6 standard. Pilot and crew modules can either be operated independently or they can be linked together allowing crews to fly a full operational mission. Both student and experienced Sea King aircrew regularly return to the simulator to undertake periodic refresher training and emergency procedures etc. The Culdrose simulator is also used by five different countries and by RAF Sea King HAR 3 crews.

The use of Full Mission Simulators has gone a long way to increase flight safety and help prepare aircrews to cope with the unexpected during high workload phases of a mission. Pilots can regularly practice their emergency procedures throughout any phase of a day or night mission. The simulator controller can progressively increase an aircraft's malfunctions during a sortie allowing the aircrew to systematically go through all the laid down procedures for those particular problems. Simulator emergency procedure training has without doubt saved many lives as pilots and aircrew are now becoming more familiar with the many combinations of aircraft malfunctions. These are now more easily anticipated as aircrews become more practised at overcoming the most demanding of situations.

One of the most unpleasant situations for an ASW Sea King crew to overcome is an AFCS malfunction, at night, in the hover and this plus other emergency procedures are regularly practiced on the RNAS Culdrose simulator. All types of emergencies can be initiated and the more regular ones include: engine bay fires, restarting an engine in flight, single engine failure, double engine failure in forward flight, computer malfunctions/freeze, runaways up and down, AFCS/control malfunctions, generator failures, fuel booster pump failure and fuel filter blockage, tail rotor drive and control failure, main gearbox malfunctions and chip warnings. As a final resort aircrew are also taught how to undertake a controlled water landing and on a positive note, how to initiate a safe water take-off.

During their time with 706 Squadron student pilots will have completed a general handling type conversion and gained their Instrument Flying Rating before moving on to Operational Flying Training (OFT).

Sea King ASW Operational Flying Training is undertaken by 810 Naval Air Squadron. Students arriving on the Squadron should have over 100 helicopter hours in their log book and will undertake a further 34 hours

Below: Sea King HAS Mk 6 (XZ580/507) from 810 Squadron based at RNAS Culdrose in formation over Somerset. *(Patrick Allen)*

on their OFT. Pilot, observer and aircrew training is integrated on the squadron and the 22-week course ends with a three-week embarked period on the Aviation Training Ship RFA *Argus* or if required HMS *Illustrious*. At the end of their training period aircrew join one of the front-line ASW squadrons.

OBSERVER/AIRCREWMEN

Royal Navy observers arrive directly from their basic officer training at Dartmouth and join 750 Squadron to undertake their Basic Observer Training. Flying in the squadron's Jetstream aircraft the 27-week course teaches the basic navigational and other skills they will require before moving on to the ASW or AEW Sea King. Recruits wanting to be aircrewmen must be experienced ratings and undergo a selection procedure for the specialisation. Both aircrewmen and observers move to 706 Naval Air Squadron where the ASW observers get their first introduction to the ASW Sea King before going to 810 Squadron for their Operational Flying Training. 849 Naval Air Squadron run advanced training courses for observers destined for the Sea King AEW Mk2. This course is one of the most expensive at almost £2 million per observer.

ROYAL NAVY HELICOPTER UNDERWATER ESCAPE TRAINING UNIT

NEW SHORT TERM AIR SUPPLY SYSTEM (STASS) COURSE FOR ROYAL NAVY AND ROYAL AIR FORCE SEA KING AIRCREW

The long-awaited Short Term Air Supply System (STASS) similar to the USN/USMC Helicopter Emergency Egress Devise System (HEEDS) was finally issued to RN helicopter aircrews in May 1992.

Front-line Commando Sea King helicopter aircrew from 846 Naval Air Squadron were amongst the first RN aircrew to complete the first Short Term Air Supply System (STASS) course at the nearby Royal Navy Helicopter Underwater Escape Training Unit (HUET) on 22 May 1992. The new STASS course involves both lectures and a practical session in the celebrated Yeovilton 'Dunker' and once completed, aircrew are then immediately issued with their own personal STASS.

Known in the US military as the HEEDS (SRU-136P Helicopter Emergency Egress Devise) it is a modified civilian SCUBA diver emergency air bottle system which entered service with the USN and USMC in 1988. It has since been responsible for the saving of a number of USN/USMC aircrew who were trapped or disorientated inside their sinking or inverted helicopter giving the extra vital minutes for their escape.

At a cost of £150.00 each, STASS is less than twelve inches long and comprises: a small stainless steel compressed air bottle filled with a mixture of 21 per cent Oxygen (02) and 79 per cent Nitrogen at a pressure of 3,000 psi/207 bars, plus an on-demand air valve/regulator, press for air 'purging' button, mouthpiece, dust cover and an air content gauge. With a service temperature limit of between +43°C to –30°C a full STASS should give approximately two minutes of air, at a depth of 20 ft at a temperature of 13°C. This time can be reduced to approximately one minute due to extremely low temperatures or by the mental/physical state of an individual and their ability to cope with the stress of a crash or immersion into a cold sea.

One of the disadvantages of STASS is the use of pressurised air and the inherent risk of air embolisms or the 'bends' as aircrew gulp in and retain pressurised air in their lungs as they head towards the surface. A Sea King can sink at the rate of 600 ft per minute which in some cases can increase this risk.

The RN STASS training programme concentrates on overcoming these problems, familiarising aircrew with the STASS and the technique of purging the mouthpiece of water prior to use and making certain that aircrew use the system correctly, expelling air as they head towards the surface. As a precaution, all students having completed the course must wait for at least ten minutes to confirm they are not suffering any abnormal symptoms prior to leaving the HUET building. A decompression chamber has been installed with fully qualified RN diver instructors on hand should there be a problem.

Issued on a personal basis, aircrew wear their new STASS in a specially fitted pocket, located on the right side of their aircrew life

Top left: The CO 846 Squadron was one of the first RN Sea King pilots to undertake the Short Term Air Supply System (STASS) course at Yeovilton in May 1992. A new 'Dunking' chair has been installed at the Helicopter Underwater Escape Training Unit (HUET) building at Yeovilton along with a decompression chamber. Aircrew are given a short familiarisation course and then have to operate the system while inverted in the 'Dunking Chair'. *(Patrick Allen)*

Centre left: The STASS bottle is approximately twelve inches long and should give two minutes of air allowing aircrew who are trapped or disorientated in a submerged helicopter the extra vital minutes to escape. The system has saved lives in the USN/USMC. *(Patrick Allen)*

Bottom left: A Sea King pilot from 846 Squadron wearing the newly issued STASS air bottle which is fitted on the right hand side of the aircrew life preserver along with his other survival kit. STASS is welcomed by almost all helicopter crews who operate helicopters over the water at night for long periods. The RN HUET unit will also be training RAF aircrew within the near future. *(Patrick Allen)*

preserver. At the present time aircrew undertake the STASS training programme in a separate 'Dunking' chair in the same pool alongside the main 'Dunker' module at Yeovilton. They must STASS requalify every two years and it is thought that in time, aircrew will undertake their normal two-year 'Dunker' training wearing STASS as part of their normal Helicopter Underwater Escape Training (HUET).

Over 900 RN aircrew are due to undertake the RN HUET STASS course by 1993 followed by 300 RAF SAR/Support Helicopter and 80 Army Air Corps aircrew. At a later stage fixed wing aircrew may also be eligible for STASS training.

For Royal Navy Sea King ASW and RAF Search and Rescue Sea King aircrew, who regularly operate over the sea, at night, at low level for long periods, STASS is a well received extra insurance should the worst happen.

Right: The latest Royal Navy ASW Sea King HAS Mk 6 from 819 Squadron. By late 1991 a programme to install an emergency lubrication system for RN Sea King main transmission/gearbox had begun with the system already being installed in a Commando HC Mk 4. Sea Kings have been known to dump all their main gearbox oil within seconds. The emergency lubrication system will give a Sea King pilot vital extra minutes to try and land before the gearbox seizes. *(RNAS Culdrose)*

Below: Sea King HAS Mk 6s from 810 Squadron. *(Patrick Allen)*

Sqdn	Role	Aircraft	Location
750	Basic Observer Training	Jetstream	Culdrose
705	Basic Flying Training (Pilots)	Gazelle HT 2	Culdrose
706	Advanced Flying Training (ASW Aircrew)	Sea King HAS 5	Culdrose
810	Operational Flying Training (ASW Aircrew)	Sea King HAS 6	Culdrose
849	Operational Flying Training (AEW Observers)	Sea King AEW 2	Culdrose

SEA KING SQUADRONS — 1991

RNAS CULDROSE (HMS *Seahawk*)

771 Sea King HAR 5	Search and Rescue
810 Sea King HAS 5/6	Anti-Submarine Operational Training
814 Sea King HAS 5	Anti-Submarine (HMS *Invincible*)
820 Sea King HAS 6	Anti-Submarine (HMS *Ark Royal*)
826 Sea King HAS 5	Anti-Submarine. In Royal Fleet Auxiliary ships or frigates
849 Sea King AEW 2	HQ Flight. Training A Flight (HMS *Invincible*) B Flight (HMS *Ark Royal*)

RNAS PORTLAND (HMS *Osprey*)

772 Sea King HC 4	Search & Rescue, support of ships working up.

RNAS YEOVILTON (HMS *Heron*)

707 Sea King HC 4	Commando Helicopter Crew Conversion and Training
845 Sea King HC 4	Front Line Commando Support
846 Sea King HC 4	Front Line Commando Support

HMS *GANNET* (Scotland)

819 Sea King HAS 5/6	Anti-Submarine. Search & Rescue.

150 Sea Kings have been delivered to the Royal Navy.

Below: Sea King HAS Mk 6 (XZ580/507) from 810 Squadron with blades folded at RNAS Culdrose. This aircraft was built as a HAS Mk 2 in 1977 then converted to HAS Mk 5 then HAS Mk 6 and had previously operated with 706 Squadron. *(Patrick Allen)*

SECTION 18

SEA KING HC Mk4 COMMANDO

Westland had already successfully developed and sold the Commando Sea King Mk2 to both Egypt and Qatar when in 1978 the Royal Navy ordered their own Commando variant — the HC Mk4.

The larger and more powerful Sea King HC Mk4 was to replace the existing fleet of Royal Navy Wessex HU Mk5s which were operated by the Royal Navy Commando helicopter squadrons at RNAS Yeovilton. Similar to the Egyptian Commando Mk2, the Royal Navy HC Mk4s retained the main rotor blade fold and tail pylon fold facility of the ASW version. The primary role of the Sea King HC Mk4 was to provide the logistics support, troop transport and utility roles in direct support of Royal Marine Commando assaults and landing operations. The aircraft had to be completely amphibious and capable of working either from Royal Navy and Royal Fleet Auxiliary (RFA) ships or land based, operating in the field, alongside the troops they would support. The primary role of the Royal Marine Commando Brigade was to protect NATO's northern flank, so much of their helicopter support would take place in Northern Norway, inside the Arctic Circle during the winter.

Today the Royal Navy operates two permanent front line Commando squadrons and, if required, they can reform further Commando Sea King squadrons (848 Naval Air Squadron, Gulf War). The primary role of the Commando Sea King squadrons is to provide a simultaneous lift of two company groups of 3 Commando Brigade, Royal Marines including their 105mm guns, ammunition stores and vehicles anywhere in the world.

The first of 41 Sea King HC Mk4s (ZA290) flew on 26 September 1979 and entered service with 846 Naval Air Squadron on 26 November 1979 and immediately deployed to Norway for winter 'Clockwork' exercises. In October 1983, 707 Naval Air Squadron became responsible for the Commando Sea King HC Mk4 conversion and training and in 1986, 845 Naval Air Squadron began to re-equip with the Westland Commando Sea King HC Mk4 giving the Royal Navy two front line Commando Sea King squadrons and allowed the Wessex HU Mk5 to retire.

As the only squadron equipped with the Commando Sea King HC Mk4 during the Falklands War 'Operation Corporate', 846 Naval Air Squadron undertook a variety of essential roles including numerous covert missions. The squadron undertook the first operational night mission using ANVIS, Night Vision Goggles (NVG) flying 736 night hours moving troops, 105mm guns, ammunition, stores and deploying and extracting Special Forces troops throughout the campaign. On 8 February 1984, 846 Naval Air Squadron were back in action, supporting British Forces in the Lebanon helping to evacuate civilians from the British Embassy in West Beirut. In 1990/91 the entire front line strength of Royal Navy Commando Sea King HC Mk4s including the reformed 848 Naval Air Squadron were back in action both at sea and on land supporting 'Operation Granby'.

SEA KING COMMANDO HC Mk4
41 a/c built

First flight on 26 September 1979 (ZA290)

ZA290 (26/ 9/79) to ZA299 (10/ 9/81)	10 a/c
ZA310 (16/ 9/81) to ZA314 (20/ 9/82)	5 a/c
ZD476 (8/12/83) to ZD480 (19/ 3/84)	5 a/c
ZD625 (18/ 4/84) to ZD627 (25/ 6/84)	3 a/c
ZE425 (10/ 9/85) to ZE428 (10/12/85)	4 a/c
ZF115 (3/ 6/86) to ZF124 (4/ 4/87)	10 a/c
ZG829 (10/ 4/89 EPTS	1 a/c
ZG820 (2/ 6/90) to ZG822 (18/ 9/90)	3 a/c

Right: 707 Squadron Sea King HC Mk4s prepare to depart RNAS Yeovilton for a flypast during the Presentation of the Colour to the Naval Air Command at HMS *Heron* by the Queen on 27 June 1991. *(Patrick Allen)*

SEA KING COMMANDO HC Mk4 SPECIFICATIONS

Crew: One or two pilots, 1 aircrewman, 27 passengers.

Dimensions: Rotor diameter 62ft (18.9m). Length 72ft 8in (22.15m). Height 16ft 10in (5.13m). Length with tail pylon folded 47ft 3in.

Engines: Two Rolls-Royce Gnome H1400-1 turboshafts, 1,600shp each (Mk12201 ECU port) (Mk12301 ECU starboard)

Fuel system: AVTUR forward tank 404 gallons, aft tank 396 gallons. Total fuel 800 gallons (pressure refuel). Gravity refuelling allows 817 gallons total.

Weight: Empty 13,500lb (6,150kg. Max: 21,400lb (9,700kg).

Underslung: 8,000lb (3,628kg). Cargo (SACRU).

Performance: Max speed 125kt (231km/h). Cruise speed 100kt (185km/h). Max range 650nm (1,200km). Max endurance 6½ hours.

Weapons: Cabin mounted 7.5mm GPMG.

Defensive Aids Suite: ALQ157 Infrared Jammers, M130 Chaff and IR decoy flare dispenser. AAR47 Missile Approach Warning (MAW) and Radar Warning Receivers.

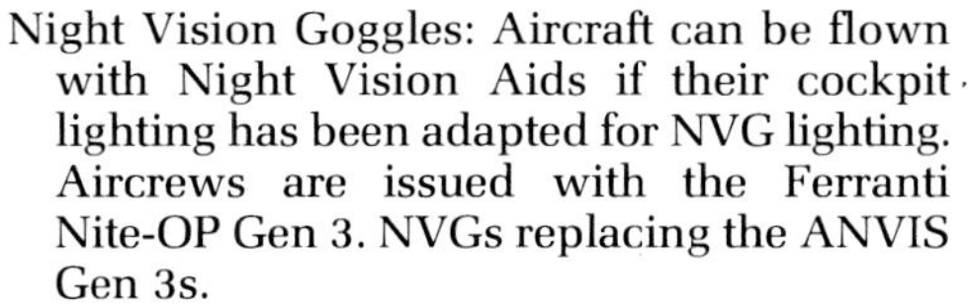

Night Vision Goggles: Aircraft can be flown with Night Vision Aids if their cockpit lighting has been adapted for NVG lighting. Aircrews are issued with the Ferranti Nite-OP Gen 3. NVGs replacing the ANVIS Gen 3s.

NBC/Chemical Warfare: Crews wear AR5 Aviator respirators.

Flight Instruments: Two Attitude Indicators, 2 APN198 Radar Altimeters, 2 Standby Attitude Indicators, 2 TANS Steering/Hover Indicators, GM Mk9 Compass System and Compass Controller, 2 Airspeed Indicators, Barometric Altimeter Mk20A, Vertical Speed Indicator, Cruise Guide Indicator, Outside Air Thermometer, Ground Speed and Drift Indicator, Standby Compass, 1 8-day Mk5A Clock.

Radios: UHF, VHF, AM/FM, HF (Secure Speech radios fitted for 'Granby'), Mode 1, 2, 3 and 4 IFF interrogation system. Submerged Aircraft Location Beacon (Projector Sonar).

Pyrotechnics: Signal pistol, Marker Marine No4 MkN9 and MkN14, Float, Smoke and Flame/Smoke Grenade 83, Chaff skillets/ cartridges.

Navigation Systems: Decca Mk19 Area Navigation System, Decca (Racal) Type 71 Doppler System, Air Data System, Decca Tactical Air Navigation System (TANS) Type 9447F/09 Mk19 Computer, Decca Automatic Chart Display, VOR/ILS (Decca RV671) Controllers, UHF (Violet Picture) and VHF Homing (DME) Distance Measuring Equipment.

1990: During the Gulf War Royal Navy Commando Helicopters were updated with a NAVSTAR Global Positioning System (GPS) linked to a Racal RNS252 (Supertans). This system replaced the Decca/Racal TANS Computer but was compatible with the existing Decca Mk19 Area Navigation System. The new TANS 252 computer can take inputs from both the GPS and Doppler System comparing and updating the aircraft's position giving information on position, height and the helicopter's velocity. The Racal 252 Computer also has a Data Transfer Devise. Large number of Waypoints stored in the system's extensive library are easily transferred to a similar TANS. This information can include Lat/Long, Bearing and Distance or grids.

The Global Positioning System has been developed over many years by the United States Department of Defense and provides a 24-hour global navigation system. The system consists of 21 space vehicles in six orbital planes with 12-hour orbits. Each satellite transmits a signal in the L-Band along with a data message which includes system clock

Left: Ten Sea King HC Mk4s formating for a massed flypast at Yeovilton Air Day 1988. *(Patrick Allen)*

correction information, ephemeris data, system message and almanac data. The small GPS receiver fitted to the top of the Sea King's tail pylon picks up these signals and this information is processed by the GPS system which also displays how many satellites are being interrogated.

Role Equipment:

Hoist/Winch: Hydraulically operated, 256ft long, maximum load 600lb.
Stretcher Installation: Six stretchers in tiers of three or six stretchers plus seating for one attendant and 10 casualties.
Ferry Tank: Single 1,500lb (184 gallons) fuel ferry tank can be fitted.
Sand Filter: A Vortex type sand filter can be fitted over the cockpit forward of the engine air intakes. The filter comprises two scavenger fan units and when an engine is running, air is drawn through the vortex separator tubes into the filter unit and any sand particles/solids are then forced against the sides of the tubes by swirl vanes. The clean air in the centre of the tubes passes into a chamber forward of the engine air intakes and the sand/solids are drawn off through exhaust ducts by the fans and ejected over the side of the aircraft. During the Gulf War and in Kurdistan the sand filters proved extremely successful and engine damage was below normal. The sand filters at the time of writing had a very limited icing clearance preventing the use of the system in the temperate zone. During the winter of 1991/92 cold weather trials took place to clear the system for permanent use by the Commando Sea Kings replacing the existing 'barn door'.
Paratrooping: Freefall parachuting 26 troops in clean fatigues, 12 with equipment. Static Line: 20 in clean fatigues or 18 with equipment in sticks of four.
Abseiling/Roping: One rope not longer than 200ft can be attached to the abseiling point on the winch/hoist or the static line monorail. Roping attachment to winch support frame.
Supply dropping: Both parachute stores (700lb), freedrop 200lb.

AIRCRAFT LIMITATIONS
RN Sea King HC Mk4

ICE AND SNOW

The aircraft is cleared for flight within the temperature range of minus 26 degrees Centigrade to plus 45 degrees Centigrade.

Top right: Having completed a Vertrep tasking this 845 Squadron Sea King returns to HMS *Intrepid* during Exercise 'Northern Wedding'. *(Patrick Allen)*

Right: 707 Squadron instructors train aircrew to operate and fire the 7.62mm GPMG cabin door gun fitted to the Commando Sea King. *(Patrick Allen)*

The cold soak limit for starting is minus 20 degrees Centigrade unless the main rotor gearbox has been pre-heated. The Herman-Nelson heater is usually used for this task.

The aircraft can be flown single pilot with a suitably trained aircrew member to carry out left-hand seat cockpit duties including manual throttles. Some operational circumstances may demand that the left hand seat is occupied at all times or that a second pilot is required, i.e. icing conditions, adverse weather or night hovering, NVG operations or prolonged sorties.

The aircraft is cleared for flight in limited engine and airframe icing conditions and is fitted with a windscreen, FOD shield, engine and intake anti-icing system comprising TKS fluid anti-icing system (five gallons AL5 anti-icing fluid) and heated elements. Engine icing conditions exist when cloud or fog reduces visibility to less than 1,000 metres, or when in rain within the temperatures plus 10 degrees to minus 30 degrees Centigrade. Operating in airframe icing conditions with composite main rotor blades exists when cloud or fog reduces visibility to less than 1,000 metres in the temperature range from zero to minus 30 degrees. Flight in freezing rain or drizzle is prohibited.

Flight in snow is permitted, but limited to the following: visibility must exceed 500 metres and the minimum outside temperature must exceed 10 degrees Centrigrade. Hovering in recirculating snow must be limited to one minute continuous and no more than ten minutes per sortie. During winter deployments to Norway IFR flight is prohibited and all Commando Sea King flights are VFR 'CLEAR OF CLOUD IN SIGHT OF THE SURFACE'.

Weight Limit: Normal maximum all-up weight for all manoeuvres is 21,000lb. The maximum overload weight is 21,400lb only to be considered in cases of operational need.

Cargo Door: Flight with the cargo door full open is restricted to 90kt maximum or with door one-third open 115kt.

Flotation Gear: Consists of two cool-gas filled inflatable bags, one on each stub wing. These can be fired either manually or by submersion actuators.

Top left: While embarked aboard HMS *Intrepid* much of the Sea King maintenance must be carried out in the open. This can be particularly demanding during rough weather or in extreme cold. *(Patrick Allen)*

Left: Rolls-Royce Gnome H-1400-1 turboshaft engine, one of two belonging to a Sea King Commando HC Mk 4. The engines develop a maximum 1,660shp and 1,250shp continuous. *(Patrick Allen)*

707 NAVAL AIR SQUADRON
RNAS Yeovilton

707 Squadron was originally commissioned in 1964 at RNAS Culdrose as the Royal Navy's advanced and operational Commando Flying Training Squadron. The squadron moved from Culdrose to Yeovilton in 1972. Originally equipped with the Wessex HU 5 it began to re-equip with the Commando Sea King HC Mk4 in October 1983.

Operating eight Sea King HC Mk4s the squadron's primary role is the advanced and operational training of pilots; pilot conversion and refresher training; aircrewmen conversion and refresher training; the training of all maintenance personnel on Sea King Mk4. Most squadron graduates pass on to the front line Commando squadrons, 845 and 846.

Royal Navy helicopter pilots undertake the same basic flying/training courses with those streamed for Commando helicopter operations leaving RNAS Culdrose after obtaining their 'Wings' with 705 Naval Air Squadron and joining 707 Squadron at RNAS Yeovilton to undertake their Advanced and Operational Flying Training. The advanced stage teaches pilots how to fly the aircraft when the operational phase teaches the various skills needed for operational flying.

The course includes: mountain flying, either in Snowdonia, Bavarian Alps or the Pyrenees mountains where they are taught basic mountain flying techniques including wind finding and take-off, landings and flying around pinnacle, ridges, spurs, valleys, bowls and emergencies etc. Other exercises to master include: low level navigation day and night, confined area operations, troop drills, close and tactical formations, day and night load lifting, wet and dry winching, day and night deck landings, fighter evasion plus a basic SAR course are all included in the syllabus. Commando Sea King pilots also make full use of the Sea King simulator at RNAS Culdrose undertaking their emergency procedures during the AFT phase. At the end of the 13-week, 64-hour OFT course, students undertake the tactical phase, which takes the form of a military exercise. Often this takes place in Germany and provides an opportunity to put into practice all the tactical skills that they have been taught before moving onto a front line squadron.

Top right: 846 Squadron undertook some hazardous missions flying 105mm guns forward during the Falkland War. This picture shows 707 Squadron Sea Kings moving 29 Commando RM, one-ton Landrovers and 105mm guns during a training mission. *(Patrick Allen)*

Right: An 845 Squadron Sea King in Arctic camouflage prior to deploying to 'Clockwork' at RNoAF Bardufoss in Norway 1989. *(Patrick Allen)*

'Clockwork'

Having joined a front-line squadron, Commando Sea King pilots continue to undertake flying training as they master the more specialised tasks undertaken by the Royal Navy Commando Squadrons. These include Arctic Warfare Training and Night Vision Goggles (NVG) operations.

Each year members of the Yeovilton-based Commando helicopter squadrons undertake winter training in northern Norway, this period of training is called 'Exercise Clockwork'.

Each winter (except for 1990/91) 845 and 846 Naval Air Squadrons detach three aircraft each and up to 70 officers and ratings to RNAoF Bardufoss, a Norwegian Air Force base situated 160nm inside the Arctic Circle at latitude 70 degrees north. Aircrews are taught Arctic survival and the flying training teaches pilots snow landing techniques as well as the finer points of mountain flying, day and night navigation, Arctic troop drills and load lifting in recirculating snow by day and night.

During Arctic operations the standard loading for Royal Marine trooping is 16 men in Arctic equipment (280lb each) with four ski bundles and two pulks @ 320lb = 5,120lb.

One of the more demanding exercises undertaken are 105mm gun lifts by day and night. These can be difficult in heavy recirculating snow and pilots are taught how to use hover reference marks (HRMs) as their external reference marks. These can be any non-moving object such as a BV206, vehicle, another load waiting to be lifted or troops. Without HRMs it is easy to become disorientated in heavy recirculating snow. NVG gun lift training is also included for the more experienced aircrews.

For the maintenance teams the period in Norway provides experience in servicing aircraft in the open at temperatures down to −30°C as well as learning to survive in the Arctic and having to fight to defend the aircraft from a tactical operating base known as 'Eagle Bases'. The 'Clockwork' training period culminates in a week's tactical camp living under canvas in the field and putting into practice the lessons learnt over the previous months' training by flying and servicing the aircraft in hostile environment. 'Clockwork' training is conducted between November and March every year and the fully trained aircrews proceed to the large NATO winter exercise operating alongside the Royal Marines in the field.

Top right: HMS *Intrepid* in Plymouth Sound 1988 with her Commando Sea Kings swarming around her. *(Patrick Allen)*

Right: Royal Marines deploy aboard 845 Squadron Sea Kings at RNAS Yeovilton. *(HMS Heron)*

SECTION 19

ROYAL NAVY COMMANDO SQUADRONS

846 NAVAL AIR SQUADRON

As the senior of the two front-line Royal Navy Commando helicopter squadrons, 846 Squadron is permanently at 72 hours notice to deploy on operational tasks worldwide. The squadron is also capable of operating independently in the field or embarked for up to sixty days.

Both 845 and 846 Naval Air Squadrons are based at RNAS Yeovilton in Somerset and their primary role is to provide eight Sea King HC Mk4s each to support 3 Commando Brigade Royal Marines, their amphibious forces and specifically on the Northern Flank in Northern Norway. To this end aircraft and aircrews spend up to four months each year operating in Arctic conditions in winter Norway. As well as being trained in Arctic warfare the squadrons are also capable of deploying worldwide at short notice. This was admirably demonstrated during the Gulf War and Kurdistan when at short notice the Commando squadrons, including a reformed 848 Squadron, deployed to the Gulf and Saudi Arabia. Having returned from the Gulf, 846 Squadron was within two weeks sent to Kurdistan.

846 Naval Air Squadron was the first Royal Navy Commando squadron to become operational with the Sea King HC Mk4 and received their first Sea King on 26 November 1979. They immediately deployed the new aircraft to Northern Norway to take part in the Arctic warfare exercise called 'Clockwork' supporting Royal Marines during their winter training. The last of the squadron's Wessex HU Mk5s were handed over to 845 Squadron on 23 October 1981 leaving 846 Squadron a total of eight Sea King HC Mk4s.

The squadron played an important role in the Falklands conflict with the entire squadron embarking with the initial task force in HMS *Hermes* with elements in HMS *Fearless*, *Intrepid*, SS *Canberra* and MV *Norland*. During the operation the Sea Kings flew a total of 2,800 hours, completed 10,000 individual troop movements and transported over 18 million pounds of freight.

Below: A pair of 846 Squadron Sea Kings arrive to collect Royal Marine troops during an exercise in Norway 1988. *(Patrick Allen)*

BEIRUT

In February 1984 the squadron was tasked to support the British Forces in Lebanon flying many missions each day over war torn Beirut. On 8 February 1984, British Forces in Beirut were ordered to give up their forward camp at Haddath. During that day the squadron airlifted 100 men and 47 underslung loads from the port of Junieth to RFA *Reliant*, the last three sorties being completed using Night Vision Goggles (ANVIS GEN 3s). Once this task was completed the order was received to evacuate British and Commonwealth civilians assembled at the British Embassy in West Beirut. On 10 February the squadron airlifted 521 civilians to RFA *Reliant*. Apart from the 191 British passport holders, a further 330 people of 31 different nationalities were brought to safety. In recognition of its contribution to the peace initiative the squadron was awarded the Boyd Trophy for 1984

WINTER NORWAY

Each winter since 1978 the squadron deploys to Norway to support 3 Commando Brigade, Royal Marines as they undertake their Arctic warfare training. In 1987 the squadron moved to a new winter location at Tretten, near Lillehammer in southern Norway. Each year the squadron detach aircraft and aircrew to both 'Clockwork' at RNoAF Bardufoss and to Tretten. During their 'Clockwork' training period at RNoAF Bardufoss new aircrew or those requiring refresher training are given experience of the problems associated with severe cold weather operations. Arctic flying training includes snow landing techniques as well as the finer points of mountain flying, day and night navigation, Arctic troop drills and load lifting in recirculating snow by day and by night. For the maintenance teams the period at 'Clockwork' provides experience in servicing aircraft in the open at temperatures down to −30°C as well as learning how to survive in the Arctic and having to fight and defend the aircraft from a tactical operating (Eagle Base) base in the field.

During the 'Clockwork' training period the remainder of the squadron is located at Tretten, supporting Royal Marine troops in both the north and south of Norway. By February the three 'Clockwork' aircraft are released to join the remainder of the squadron at Tretten, until the final winter exercise period in March. During their period at Tretten the squadron undertake numerous

Top left: Arctic camouflage painted Sea King (VP) from 846 Squadron being refuelled in the mountains. The arctic paint has proved ideal at concealing the Sea Kings from being seen by fast jets helping to reduce the risk from an air attack while flying or on the ground. *(Patrick Allen)*

Left: 846 Squadron Sea Kings (VP/VZ) load up with Royal Marines at Bessheim (Ice Station Zebra) during an exercise in Norway in 1990. *(Patrick Allen)*

different types of taskings in support of the Royal Marines as they undertake their Arctic warfare training (AWT).

Tasking includes first and last light insertions of the Mountain and Arctic Warfare Cadre (M&AW), paratrooping, troop moves by day and at night using NVGs. The squadron also detach Flights to several training areas including the mountain artillery ranges where they lift the 105mm howitzers of 29 Commando Royal Marines both by day and at night using Night Vision Goggles (NVG). Aircrew undertake a squadron NVG training programme back at RNAS Yeovilton and once Arctic trained they undertake winter NVG training with the squadron whilst at Tretten.

SQUADRON NVG TRAINING

The basic NVG course comprises nine hours NVG flying training in the UK, plus one and a half hours NVG flying per month to remain current, plus normal night flying. Having qualified in the UK and gained experience as a second pilot on NVG taskings they can then progress to Arctic NVG flying. Arctic NVG training comprises three hours NVG flying in Norway learning navigation techniques and how to undertake normal snow and recirculating snow landings and take-off techniques. NVG snow flying takes experience as the goggles cannot see through snow and with limited monocular vision multiple hover reference markings (HRMs) must be used to gain a proper outside reference. White-Out is a real problem for the non-experienced pilot and it is easy to lose sight of the surface and become disorientated. Flying above the tree line in Norway is notoriously difficult and the risk of white-out and disorientating is high. Pilots must keep ground references at all times and when wearing NVG the monocular vision makes this type of flying even more demanding. In 1990-91 the squadron exchanged their ANVIS GEN3 NVGs for the new Ferranti NITE-OP GEN3s.

During the Tretten detachment the squadron keeps an aircraft ready for any local SAR taskings. In the past these have included flying injured and sick soldiers out of the mountains and to the nearest hospital. They have made good use of their NVG experience to find lost and injured Norwegian civilians in the region and they also exercise each winter with the local Norwegian Avalanche Rescue Dog Teams (Norske Lavinehunder). At the end of the winter period the squadron usually embark for a short period before deploying ashore in northern Norway to support the final exercise.

Top right: Five Commando Sea Kings aboard HMS *Invincible* during winter exercises in northern Norway. *(Royal Navy)*

Right: A pair of 846 Squadron Sea Kings deploy Royal Marines from 42 Commando in severe arctic conditions on a mountain in Norway during the winter of 1989. *(Patrick Allen)*

In October 1987 the squadron was once again involved in out of area operations. 846 Squadron was called up to provide an aircraft to support mine sweeping operations in the Arabian Gulf during 'Operation Cimnel'. Based firstly on RFA *Tidespring* and later on RFA *Olna*, the deployment lasted until June 1988. Annually the squadron is committed to numerous NATO exercises in Northern Europe, the Mediterranean and around the world. 846 Squadron was presented with the Australia Shield for 1990 which is awarded to the most operationally prepared squadron in the Fleet Air Arm. During 1991 the squadron deployed to the Gulf, Kurdistan and in November the first elements of the squadron left for 'Clockwork' training in Norway before the remainder of the squadron departed for another winter in Norway and their participation in Exercise 'Teamwork 92'.

Below: With her deck full an 846 Squadron Sea King winches down priority stores to the deck of HMS *Intrepid* during an exercise in the fjords of northern Norway. *(Patrick Allen)*

COMMANDO SQUADRONS

The primary role of both Commando squadrons is to support the amphibious forces assigned to protect the Northern Flank of NATO. The flexibility of the Commando Sea King squadrons also enables them to deploy worldwide in support of the Royal Marines. The primary task of the squadrons is to provide a simultaneous lift of two company groups of 3 Commando Brigade Royal Marines, including their 105 mm guns, ammunition and vehicles.

Both squadrons provide the tactical air mobility for the Royal Marines, supporting them wherever they may be. Regular exercises see the squadron aircraft in Norway, Germany and the Mediterranean, and when embarked in Royal Navy ships as far afield as the Caribbean and the Middle East.

To be able to maintain the rapid availability required by the Royal Marines, the Royal Navy Commando Helicopter squadrons deploy and live in the field alongside the troops they support. These field locations are called Forward Operation Bases and 'Eagle Bases'.

EAGLE BASES

For field or wartime operations the squadron is split into two Forward Operating Bases (FOBs). Each FOB consists of one Command Post (CP) and four 'Eagle Bases' each with one Sea King helicopter. The Command Post (CP) receives taskings from the Brigade via CHOSC (Commando Helicopter Operations Support Cell). CHOSC allows all the Brigade's helicopter requests to be viewed through a single body, allowing a smoother more efficient use of helicopter assets. The CP, once they receive their tasking, can then choose the most available aircraft from an Eagle Base.

Each Eagle Base supports one aircraft and has a complement of ten men. This team comprises two pilots, an aircrewman, a tent leader and a mixture of junior and senior ratings with weapons/electrical, radio and mechanical skills. All forward operating bases (FOBS) have a Royal Marine to ensure that individual Eagle Bases are up to standard on field skills such as concealment, defence and sentry routines etc.

Helicopter and team operate independently from their Eagle Base location, and maintain contact with the Command Post by a secure landline telephone and by a daily 'O' Group meeting. These autonomous Eagle Bases are constantly moved to protect them from air attack or surveillance. FOBs can be moved and set up again within a very short time.

Each Sea King can move its own Eagle Base team in one move or can continue with a task and collect its Eagle Base team at a later time. The Eagle Base system is highly flexible and has proved extremely effective in increasing aircraft availability in all theatres of operations from the Arctic to the Desert.

SECTION 20

OPERATIONS WITH 846 NAVAL AIR SQUADRON

AWARD OF AUSTRALIA SHIELD TO 846 NAVAL AIR SQUADRON

During a ceremony at RNAS Yeovilton in December 1991, Rear Admiral Colin Cooke-Priest, the Flag Officer Naval Aviation, presented the Australia Shield for 1990 to 846 Naval Air Squadron.

1990 started with the squadron deploying to southern Norway supporting Royal Marines undergoing Arctic training. This was followed by embarkation on the aircraft carrier HMS *Invincible* for winter exercises during which they undertook numerous search and rescue flights for the Norwegian Red Cross Mountain Rescue Teams, often in appalling weather conditions. Straight from the Arctic 846 then exercised in Egypt, Sardinia and from naval ships in the Mediterranean. Their return to the UK coincided with the invasion of Kuwait and part of the squadron ('B' Flight) was sent to the Gulf on the support ship RFA *Fort Grange*, whilst the remainder of the squadron returned to Norway for another exercise. Immediately on return from Norway the rest of 846 deployed to the Gulf on the casualty evacuation ship RFA *Argus*. Whilst in the Gulf the squadron pioneered the development of new flying techniques whilst wearing full chemical protection clothing (AR5s) and Night Vision Goggles operating from darkened sites and from the ship.

In the Gulf they developed a reputation among the multi-national naval force for exceptional availability, delivering anything, anywhere at any time, on time. This typifies the impressive level of operational readiness achieved by the squadron and for which the award was made.

A comment heard at Yeovilton during the Gulf period sums up the period: 'Trained for the Arctic, dressed for the jungle, sent to the desert!''

APPENDIX

Like other military units throughout the Western World, the Royal Navy Sea King squadrons have recently experienced a number of changes. These changes are likely to continue into the foreseeable future. A number of significant changes have still to be announced and it is difficult, with any degree of accuracy, to predict how these will ultimately affect the Sea King squadrons.

The Royal Navy Commando squadrons are assured a busy future supporting Royal Marine Commando forces as they undertake more specialised and demanding roles as part of the UK's smaller but more mobile military force. The mobility and flexibility of the helicopter is becoming increasingly important in these new roles.

Emphasis is now being placed on the ability to provide a global rapid reaction force which is also capable of operating 'out-of-area'. New areas of interest now include NATO's Southern Flank, the Middle East as well as further afield. The ability of the Royal Marines to conduct amphibious operations will continue to be important and this will be greatly improved when the new 6 Spot Aviation Support Ship/Landing Platform Helicopter (LPH) joins the amphibious fleet.

Below: Royal Berkshire Yeomanry troops including females are flown by 846 Squadron Sea Kings during an Army exercise in January 1992. *(Patrick Allen)*

Recent changes to the Commando Sea King squadrons have included both 845 and 846 Squadrons losing one aircraft, reducing their size to seven Sea Kings each. This reduction has given 772 Squadron based at RNAS Portland an extra aircraft increasing their size to six Sea King HC Mk4s. 707 Squadron continues to operate eight sea King HC Mk4s in the training role. With the possible closure of the Portland Naval Base and with FOST moving to Plymouth, there is the possibility of 772 Squadron being disbanded or being moved and converted to the Sea King Mk5 allowing their Sea King HC Mk4s to return to the Yeovilton fleet.

These changes have not reduced the Royal Navy's ability to support the Royal Marines. The Commando squadrons continue to provide the Royal Marine Commando forces with 24 Sea King HC Mk4s. During time of war, both 845 and 846 Squadrons will increase their strength to 12 aircraft, taking additional Sea Kings from 707 Squadrons and from those held in reserve. Today, there are 36 Sea King HC Mk 4s in the Fleet. Those not operating with either 707, 772, 845 and 846 Squadrons are either undergoing deep maintenance or being held in reserve.

By the summer of 1992 all the Commando Sea King fleet had been overhauled and updated with the latest operational equipment first fitted during the Gulf War. Although all this equipment had originally increased the aircraft's weight, the removal of the original Decca navigation equipment (now replaced by GPS) and the replacing of old radios with the latest lightweight 'Have Quick' radios has helped reduce aircraft weight.

One unforeseen benefit of the Gulf War was the need for all the Commando Sea Kings to be overhauled, stripped down and repainted. Over the years the Sea Kings had been re-sprayed a number of times and some were carrying several coats of green paint. Once stripped down and repainted they were over 200 lb lighter!

The Sea King HC Mk4 will soldier on well into the next century. At the present time there has been no decision regarding a Commando Sea King replacement. The Black Hawk is a popular choice with the Squadrons, although a stripped out EH101 Merlin with radar, NVG compatible cockpit and rear loading ramp is also a possibility.

Below: 707 Squadron and 846 Squadron Sea Kings are seen lifting one-ton Landrovers in support of 29 Commando, Royal Marines. Twelve Sea King HC Mk 4s took part in the Yeovilton Air Day 1992. *(Patrick Allen)*

Bottom: Troops run to board an 846 Squadron Sea King (ZA291/VN) during an early morning troop move on Salisbury Plain in January 1992. *(Patrick Allen)*

Left: 846 Squadron Sea King HC Mk 4 (ZG821/VL) transits around Salisbury Plain prior to dropping off 16 troops from 40 Commando, Royal Marines during Exercise 'Rolling Deep 92'. *(Patrick Allen)*

Below: Charlie Company, 40 Commando prepare to 'hover jump' from an 846 Squadron Sea King during Exercise 'Rolling Deep 92'. *(Patrick Allen)*

Right: Exercise 'Rolling Deep 92' saw the first trial underslung load lifting by an 846 Squadron Sea King (ZA314/VK) of the new 'SAKA' Light Strike Vehicle (LSV) on trial with the Royal Marines for the reconnaissance role. LSVs are used by a number of specialist units for various covert missions. *(Patrick Allen)*

Below: ZA292/VH was one of the original four Sea King HC Mk 4s built by Westland (WA872) and first flew on 20 December 1979. Operated by 846 Squadron 'VH' is seen fitted with a low response beam carrying a 3½ ton Volvo Bandvagn over-snow vehicle in northern Norway. *(Westland)*

SECTION 21

FALKLAND CONFLICT

Over 60 Westland Sea Kings deployed to the South Atlantic on 'Operation Corporate'. Land and sea based Sea Kings undertook a record 11,922 flying hours totalling 5,552 sorties including 8,613 deck landings within the space of three months.

The Sea King soon became the workhorse on land and sea, protecting the Task Force and providing logistical and troop lifting support ashore. In their Combat SAR role, they helped to recover aircrew from the sea and survivors from damaged ships which included HMS *Sheffield*, *Coventry*, *Antelope* and RFA *Sir Galahad*. As the main protectors of the Task Force one ASW squadron (826 Sqn) of nine Sea Kings kept two of their aircraft on station continuously for 30 days with seven Sea Kings from another squadron moving over 450 tonnes of stores plus 520 troops to unprepared landing sites in a single day. The conflict saw the first operational use of the HIFR refuelling method in combat allowing an 826 Naval Air Squadron Sea King to fly continuously for a record 10 hours 20 minutes. Many of the Sea King crews recorded over 100 flying hours per month in their Log Books with one Sea King clocking up 265 hours in a single month.

During the conflict, 846 Naval Air Squadron undertook the first operational covert night missions using ANVIS Night Vision Goggles (NVG) and totalled over 736 night flying hours supporting Special Forces (SAS/SBS) operations, moving troops, 105mm guns, ammunition, stores and Rapier missiles. Throughout the campaign, Sea King availability both on land and at sea remained in excess of 90 per cent, a tribute to the helicopter's strength, reliability and the skill of their Navy maintainers who had to operate in harsh combat conditions.

820 NAVAL AIR SQUADRON

820 Naval Air Squadron deployed nine Sea King HAS Mk5s aboard HMS *Invincible* providing the ASW and Surface Search for the Task Force. One of the squadron's Sea Kings was always airborne on Screen and once inside the Total Exclusion Zone (TEZ) the squadron often had three aircraft airborne on ASW operations plus another Sea King on surface search and missile decoy operations. The squadron undertook numerous other types of missions including SAR, HDS and Vertrep. During 'Operation Corporate' the Squadron flew 2,150 hours undertaking over 700 sorties.

824 NAVAL AIR SQUADRON

Equipped with six Sea King HAS Mk2As the Squadron was split into three Flights. Two flights 'A' and 'C' deployed aboard RFA ships in the South Atlantic while 'B' Flight deployed to Gibraltar.

'A' Flight deployed aboard the Fleet Tanker RFA *Olmeda* with two Sea King HAS Mk2As undertaking ASW, HDS and Vertrep missions both at sea and ashore in the Falkland Islands, before deploying down to Southern Thule, South Georgia for 'Operation Keyhole'. The Flight recorded a single HDS mission lasting 9 hours 30 minutes.

'C' Flight with three Sea King HAS Mk2As deployed aboard the Fleet Replenishment ship the RFA *Fort Grange* to provide the Helicopter Delivery Service (HDS) to the 25 vessels of the Task Force. Sea King HAS 2A (XV699) was lost on 11 July due to an engine failure and subsequent ditching. During the Conflict the Flight recorded moving over 2,000 tonnes of stores in 650 hours of transfer operations.

'D' Flight — Designed and developed within a record two months, the first flight of the Westland Sea King AEW2 took place on 23 July 1982. Two modified Sea King AEW2s (XV704/XV650) joined 'D Flight' 824 Naval Air Squadron who had reformed on 14 June 1982 and deployed to the South Atlantic aboard HMS *Illustrious* on 2 August 1982. The Flight returned to RNAS Culdrose on 7 December 1982 and on 1 November 1984 the Sea King AEW Mk2s joined the reformed 849 Naval Air Squadron.

825 NAVAL AIR SQUADRON

825 Naval Air Squadron reformed on 3 May 1982 and deployed to the South Atlantic with 10 Sea King HAS Mk2/2As. All the ASW and Sonar equipment was removed from the aircraft and trooping seats fitted. The helicopters were deployed to bolster the Commando Sea Kings from 846 Squadron and to undertake the trooping and logistical lift role. The squadron deployed ashore as soon as possible and supported the ground forces as they retook the Island. The Sea Kings moved troops, stores, ammunition and undertook Casevac missions as well as moving Rapier missile batteries and helping the Commando Sea Kings during the 105mm gun and ammunition lifts. 825 Squadron Sea Kings helped in the recovery of casualties from the RFA *Sir Galahad* and RFA *Sir Tristram* and in a period of two weeks flew over 1,700 hours.

826 NAVAL AIR SQUADRON

Equipped with nine Sea King HAS Mk5s, 826 Squadron embarked on HMS *Hermes* to undertake the ASW and Surface Search roles. Three of the squadron's Sea Kings maintained an almost continual ASW screen 12 miles ahead of the Task Force with a fourth Sea King on Surface Search duty up to 200 miles ahead. The squadron Sea Kings undertook the first HIFR operations in action with one Sea King airborne for a total of 10 hours 20 minutes. As well as undertaking their ASW/ASV missions the squadron undertook numerous SAR missions. The squadron recorded maintaining three aircraft in the air around the clock for two months flying over 3,000 hours.

846 NAVAL AIR SQUADRON

846 Squadron deployed to the South Atlantic with twelve Sea King HC Mk4s. Nine aircraft deployed aboard HMS *Hermes* and three Sea Kings aboard HMS *Fearless*. One additional Sea King was airfreighted to Ascension Island to help with the movement of stores and another deployed aboard HMS *Intrepid* as she sailed past RNAS Culdrose bringing the squadron total to 14 Sea King HC Mk4s. Once the Task Force arrived at Ascension Island 846 Squadron aircraft were redeployed around the fleet embarking aboard SS *Canberra*, MV *Elk*, MV *Norland*, HMS *Hermes*, HMS *Fearless* and HMS *Intrepid*.

Five of the squadron's Sea Kings were embarked aboard HMS *Hermes* with four of these aircraft fitted with Night Vision Goggle (NVG) compatible cockpit lighting and assigned to Special Forces missions. The day before the squadron left the UK, they were issued with the recently developed (Aviator Night Vision System) ANVIS (GEN 2/3) Night Vision Goggles (NVG) which had only just arrived from the USA. There was no time to train any of the crews in NVG flying techniques prior to leaving the UK and the squadron experimented as best they could with the new goggles on the transit down to the Falklands. With only a brief stop at Ascension Island which was then cut short due to a hostile submarine contact, there was no opportunity to gain any overland NVG experience prior to the first operational missions. During the transit south, the squadron lost their first Sea King (ZA311/VP) on 23 April as it undertook a night Vertrep mission.

The four Special Forces (NVG) Sea King HC Mk4s undertook their first covert mission on 1 May flying SAS/SBS troops ashore while the Task Force was still 200 miles out. These Special Forces troops not only had to be covertly inserted on the islands, but they also needed to be regularly re-supplied and extracted. Having been issued with a limited number of ANVIS Generation 2 and improved Generation 3 NVGs, these covert missions were always flown with two pilots plus a crewman. The flying pilot would wear the later, improved Generation 3 (GEN3) goggles with the second pilot and crewman wearing the earlier Generation 2 (GEN2) goggles. Many of these missions were flown in poor weather conditions with little or no ambient light. This often led to the less powerful GEN2 goggles, which required more ambient light to close down, leaving the pilot to struggle along on his own.

Without Decca navigation in the South Atlantic, the Sea Kings needed to rely on their Doppler Tactical Air Navigation System (TANS) for these long range transits. Prior to departing from HMS *Hermes* on one of these

Below: 846 Squadron Sea King HC Mk 4 on an HDS mission to RFA *Sir Geraint* on the passage south. HDS and VERTREP missions were a regular feature for Royal Navy Sea King crews during the War. *(Flight Deck/FAA)*

Bottom: 826 Squadron Sea King HAS Mk 5 aboard HMS *Hermes*. The squadron undertook almost continuous ASW and surface search missions during the conflict protecting the Task Force. The squadron aircraft amassed over 3,000 flying hours during the deployment. *(Flight Deck/FAA)*

missions, the crews would undertake a full mission brief several hours before departure and run through the entire sortie from beginning to end. It was vital that the helicopters' navigational systems had the precise position of the ship on departure and considerable time was spent before each mission synchronising and updating the Sea King's, TANS to match the ship's exact position. This initial FIX needed to be extremely accurate as it would be used to steer the aircraft to the Falkland Islands and more importantly to guide them back to the ship's estimated position on their return.

Often departing HMS *Hermes* at around two or three a.m. in the morning these long range covert missions often lasted over four hours. Flying on NVG at heights below 50 feet, usually in a formation of three aircraft, the Sea Kings would approach East Falkland Island from the north and then visually locate Concordia Rock on the north west of the Island as a precise Waypoint, update their TANS, then split up and transit inland to their individual destinations. Careful pre-mission planning was taken to achieve suitable concealed approach and departure (CADS) routes through the Islands so as not to compromise the helicopters and their customers. The squadron undertook these Special Forces missions almost every other night often flying extremely long distances.

The Special Forces Sea Kings were also used to insert SAS/SBS troops during the Goose Green and Pebble Island raids moving at one point over 45 troops in two aircraft. During the Pebble Island Raid on the night of 11–12 May, HMS *Hermes* sped towards the Falkland Islands to allow two of the squadron's Sea Kings to launch at close range for the mission. A high sea and the reluctance of the ship to reduce speed in a potentially dangerous location, prevented the Sea Kings from spreading their main rotor blades. Within 10 minutes of the sortie being cancelled the two *Hermes*-based Sea Kings loaded with SAS troops from 'D' Squadron were able to lift-off and were then joined by another of the squadron's Sea Kings to complete the mission.

The SAS destroyed 11 Argentine aircraft on Pebble Island. During this early phase of the conflict the remaining 846 Squadron Sea Kings were also kept busy moving 105mm guns, troops and undertaking numerous Vertrep missions.

On 18 May 1982 HMS *Hermes* undertook another of her high speed dashes towards the west and launched Sea King HC Mk4 (ZA290/VC) on a one-way Special Forces mission to the mainland of South America. This aircraft, which was the first Royal Navy Commando Sea King to be built by Westland, was subsequently burnt out on a beach near

Punta Arenas in Chile. Although the mission remains secret, the fact that the aircraft arrived on the mainland and was able to undertake such a mission, considering the fuel and the weight of the aircraft, was a tribute to both the aircrew and to the maritime qualities of the Sea King. The squadron lost their third aircraft on 19 May 1982 when Sea King HC Mk4 (ZA294/VT) ditched into the sea while moving SAS/SBS troops to HMS *Intrepid*.

On 20 May prior to the landings, 846 Squadron undertook 'Operation Tornado' which involved the squadron undertaking decoy insertions of troops north of Goose Green. This was to confuse the Argentinians and make them think that Goose Green had been the area selected as the main landing area. This was also the first occasion that the NVG crews had seen the Falklands in daylight!

As soon as possible after the landings the squadron deployed ashore to 'Eagle Bases' or Forward Operating Bases (FOBS) located around San Carlos Waters. The four Special Forces Sea Kings deployed to Pollock's Passage and continued to undertake their covert missions supporting the SAS/SBS observation teams scattered around the islands, while the remainder of the squadron continued with their logistic support and troop lifting roles.

During the final battles 846 Squadron Sea Kings undertook a night lift of 105mm guns plus ammunition to Mount Kent, moved Rapier missile batteries and re-supplied troops throughout the final battle. Often flying nine hours per day, the squadron maintainers managed to keep the aircraft operational repairing one Sea King which was damaged by a tail rotor strike and another which had suffered a 20mm cannon shell through a main rotor blade.

During the campaign 846 Squadron Sea Kings totalled 3,107 flying hours on 1,818 sorties with the four NVG Sea Kings flying over 736 covert night missions. Average pilot flying hours on the squadron totalled 228 with many of these flown as single pilot except for the NVG missions which were always flown by two pilots.

SEA KINGS LOST
'OPERATION CORPORATE'

Sea King HAS Mk5 ZA132 826 Sqn — 12/5/82 (Engine failure)
Sea King HAS Mk5 XZ573 826 Sqn — 18/5/82 (Rad Alt/AFCS) (malfunction)
Sea King HC Mk4 ZA311/VP 846 Sqn — 23/4/82 (Ditched)
Sea King HC Mk4 ZA290/VC 846 Sqn — 18/5/82 (Burnt in Chile)
Sea King HC Mk4 ZA294/VT 846 Sqn — 19/5/82 (Ditched)

Below: 846 Squadron maintainers work on a Sea King aboard HMS *Intrepid* while a captured Argentine UH-1H 'Huey' lifts away. The Royal Navy's Flexible Maintenance Programme helped to keep the Sea Kings operationally available. The system allows routine maintenance/servicing to be tailored to meet operational requirements by either bringing forward or delaying maintenance schedules on individual components. *(Flight Deck/FAA)*

Below: 826 Squadron Sea King HAS Mk 5s and Commando Sea Kings from 846 Squadron aboard HMS *Hermes* in the background with an 846 Squadron Sea King, blades folded aboard HMS *Intrepid*. Sea Kings spent many hours moving stores, ammunition, supplies and troops around the Task Force in preparation for the landings. *(Flight Deck/FAA)*

Left: Troops deploy aboard an 846 Squadron Sea King to be inserted ashore during the land phase of the war. *(Flight Deck/FAA)*

Below: Royal Navy Sea King helicopters undertook numerous SAR missions during the Falkland War including recovering casualties from Task Force ships which had been attacked. By the end of the 21 May 1982, HMS *Argent* was sinking and HMS *Argonaut*, *Brilliant*, *Antrim* and *Broadsword* were damaged. The following day HMS *Coventry* was hit and on 25 May SS *Atlantic Conveyor* was added to the list. The value of Sea King helicopters in their SAR/CASEVAC role culminated in the RFA *Sir Galahad* rescue on 8 June 1982. HMS *Antelope* was another casualty to air attack. *(Flight Deck/FAA)*

SECTION 22

'OPERATION GRANBY' SEA KINGS

The first Sea Kings were deployed to the Gulf in July–August 1990 when both 846 and 826 Naval Air Squadrons embarked on Royal Fleet Auxiliary ships. On 22 August, 'A' Flight, 846 Squadron detached two Sea King Commando HC Mk4s to the Fleet Replenishment Ship, RFA *Fort Grange* providing the Helicopter Delivery Service (HDS).

At the same time 'D' Flight 826 Squadron who are based at RNAS Culdrose deployed two modified Sea King HAS Mk5s and 50 personnel aboard the Fleet Tanker RFA *Olna*. These modified Sea Kings operated as part of the multi-national team enforcing sanctions against Iraq, mainly by searching for and questioning merchant shipping. 'D' Flight 826 Squadron was replaced in December by 'C' Flight, who embarked aboard the Dutch RFA, HNLMS *Zuiderkruis* until mid-January when they moved to RFA's *Sir Galahad* and *Argus* primarily in the anti-mine hunting role. Fitted with a full defensive aids suite, forward looking infra-red camera (FLIR) and 'Demon' camera mine hunting video system, the Sea King HAS Mk5s proved invaluable in helping to locate and clear mines from Kuwaiti ports. There were estimate to be over 1,200 mines deployed by Iraq in 10 discreet crescent-shaped areas around the Kuwaiti coastline.

GULF CASEVAC SEA KINGS

During the early autumn it became obvious that Iraq would have to be forced out of Kuwait. A major conflict appeared almost inevitable. Naval activity in the Gulf escalated as sanctions against Iraq started to bite. With the increased risk to shipping, and the possibility of some future land action, the Royal Navy Fleet Medical Officer was given the task of finding a suitable vessel which could be quickly converted into a medical reception ship.

The ship needed to be able to accommodate both a hospital plus staff and patients. It should also have sufficient deck space to operate a number of Sea King helicopters. Surgeon Commander Paxton Dewar chose the Royal Navy's Aviation Training Ship the RFA *Argus*. Formerly a container ship, she was purchased by the Ministry of Defence in 1988 and converted to an Aviation Training Ship with a large five-spot deck, capable of operating all the Royal Navy's helicopters. This included the Royal Navy's Sea King replacement the EH101/ Merlin. The deck is also capable of operating both Sea Harriers and RAF Chinook helicopters.

Argus proved ideal for this latest conversion with her twin aircraft lifts plus two large hangars. There was sufficient space to convert one hangar into a two-storey 100-bed hospital. The remaining hangar and lift could be retained for use by her helicopters. Within weeks the hospital with airlocks, air filtration units, resuscitation bays, operating theatres and intensive care facilities was completed. It could even boast an independent power supply and separate 'Citadel' positive air pressure to prevent outside air contamination

Below: Sunday 28 October 1990 and four 846 Squadron Sea Kings are lined up at RNAS Yeovilton ready for their flight to Plymouth to join *RFA Argus* for 'Operation Granby'. *(Patrick Allen)*

and provide full NBC protection. The ship was also large enough to accommodate the extra personnel which would increase the ship's complement from 160 to 380.

Back at RNAS Yeovilton, 846 Naval Air Squadron, who as the senior of the two front-line (845/846) Commando squadrons, are at permanent 72-hours notice were told to prepare for deployment aboard the RFA *Argus* in the Maritime Casevac role. On 4 October the squadron received a preliminary departure date and on 15 October the first Sea King (VZ) was painted in desert camouflage. On Sunday 28 October 1990 four 846 Squadron Sea Kings (ZA293, ZA296, ZD476 and ZD478 coded VP, VK, VJ, VM) departed their home base at RNAS Yeovilton and flew in formation to RFA *Argus* lying at anchor in Plymouth Sound. She set sail for the Gulf on Wednesday 31 October having returned briefly to repair her steering gear.

PRIMARY CASUALTY RECEIVING SHIP (PCRS)

With her new medical facility and 846 Squadron Sea Kings RFA *Argus* was not classed as a Hospital Ship. Her intended role in the Gulf was to provide instant Maritime Medical Assistance, either from casualties arriving by her helicopters from either land or sea based action, or brought to her by other helicopters. She was deliberately deployed to the Gulf classed as a Primary Casualty Receiving Ship (PCRS) and as such did not need to comply with any Geneva Convention specified for Hospital Ships. Any casualties, no matter how slightly injured, who arrive on a designated hospital ship can only be returned to their units via a neutral country.

The RFA *Argus* was solely a Royal Navy asset and therefore had no red cross painted on her side or on her helicopters. She did not however, benefit from any immunity given to hospital ships etc. Her helicopters had the flexibility to undertake a variety of roles, and the ship was capable of providing the best battlefield medical facilities available anywhere in the region. The ship, and her helicopters, were also available to undertake large-scale civilian refugee evacuation if required. 846 Squadron already obtained invaluable experience in this role during their involvement in Beirut in February 1984.

'GRANBY' SEA KING ENHANCEMENTS

RFA *Argus* arrived on station in the Gulf on 15 November. 846 Squadron spent much of their transit time working crews up to operational readiness. The squadron had just received newly arrived operational enhancement equipment which had to be fitted to the helicopters. Pilots had little time to learn both the capabilities of the ship and how best to use the newly delivered enhancements.

This equipment included the fitting of sand filters and NAVSTAR Global Positioning System (GPS) linked to the new Tactical Air Navigation System (TANS) known as the Racal RNS-252 (Supertans). This enhanced accuracy Doppler/Satellite navigation system proved exceptional for both the sea and land based Sea Kings in the featureless deserts of Saudi Arabia. Using the Supertans 2 'Data Transfer Devise' large numbers of Waypoints stored in the system's extensive library are easily transferred to a similar TANS. This information can include Lat/Long, Bearing and Distance or Grids. Also fitted was a comprehensive Defence Aids Suite which included Missile and Radar Warning Receivers (RWR), ALQ 157 Infra Red Jammers, improved communications and IFF equipment, plus the fitting of 7.62mm door guns. All the crews required work-up on this equipment, and practice in day and night deck landings plus night vision goggle flying techniques at sea.

846 Naval Air Squadron's complement of Sea King helicopters normally numbers eight. These are split into four flights of two helicopters. Each flight has a complement of six pilots and four aircrew plus their helicopter maintainers. For 'Operation Granby' 'C' and 'D' Flights were embarked on RFA *Argus*, 'A' and 'B' Flights on RFA *Fort Grange*. 'A' Flight later became part of the reformed 848 Naval Air Squadron and deployed into the desert to support the 1st (British) Armoured Division on 6 December 1990.

Bottom: 846 Squadron Sea Kings on patrol in the Gulf with RFA *Argus* in the background. *(Patrick Allen)*

Below: A Royal Navy Sea King pilot wearing his AR5 (NBC) aircrew respirator and the latest Nite-Op Night Vision Goggles (NVG) prior to leaving for the Gulf. *(Patrick Allen)*

The most important part of the *Argus* aspects during the work-up period. After With up to four helicopters on the deck at

out on the Commando version. They do have a Radio Altimeter (Radalt) linked into the height hold. With the possibility of recovering survivors from the sea, crews needed to be capable of operating in all scenarios whilst wearing NVGs. During the work-up phase helicopters practised making approaches down to cyelumes thrown into the sea. These luminescent green coloured light sticks were ideal for these early exercises. Using the Radalt and height hold together, crews could reduce their height slowly by winding down the Radalt with the pilot using the cyelume as a hover reference mark. This training proved invaluable with all the crews capable of coming down to a hover over the sea while on NVG. If a ship is burning or there is enough external illumination by fires, crews can flick off NVG to normal vision and then resume NVG flying at a later time.

Left: 'B' Flight 846 Squadron Sea King ZE118/VL returning from an HDS mission high up in the Gulf returns to RFA *Fort Grange* protected by a Javelin missile team, January 1991. *(Patrick Allen)*

Below: 846 Squadron Sea King ZE425 and 848 Squadron Sea King ZG821/WN prepare to depart RFA *Argus* on their return from the Gulf in April 1991. *(Patrick Allen)*

DESERT OPERATIONS

Although based on *Argus* there was the possibility that helicopters would either be deployed ashore, or required to recover casualties from desert locations. Desert refresher training was considered essential and this was undertaken on several occasions. Each Flight managed to deploy ashore and complete both day and night NVG desert operations.

846 Squadron were no strangers to the Gulf region and to desert operations. In the spring of 1990 the squadron deployed several helicopters to Egypt for a major exercise. The priority of the Gulf training detachments was to undertake refresher flying training including NVG operations. Helicopter navigation in the desert is notoriously difficult with few way-points or landmarks. It was during desert operations that the GPS system proved invaluable. Desert flying techniques were

Right: 848 Squadron Sea King ZE428/WL seen fitted with sand filters and Desert Storm invasion markings. *(Patrick Allen)*

Below: 846 Squadron Sea King (ZA296/VK) in company with an RN Lynx in Kuwait City. *(Royal Navy)*

also practised including 'Zero-Zero' landings and operating in heavy recirculating sand. 'Zero-Zero' landing techniques require practise to consistently land-on at zero height with zero ground speed. This allows any recirculating sand from the rotor downwash to be kept behind the helicopter until the last possible moment. Rotor downwash can produce 'Brown-Out' conditions (recirculating sand) similar to flying in heavy recirculating snow.

The squadron annually deployed to Arctic Norway to undertake winter flying training, hence it was familiar with these types of flying conditions. NVG flying in the desert needs constant practice. There is little or no depth perception and refresher training is required to operate safely at low level. Also included in the training programme were multiple helicopter operations where accurate positioning and NVG formation flying is particularly demanding. By the start of 'Desert Storm' the squadron was capable of operating from land or sea and had the flexibility to undertake numerous specialist tasks.

Thankfully, RFA *Argus* was not put to the test regarding large-scale casualty evacuations. The ship and her helicopters were however, kept busy undertaking numerous other tasks. Between 16 January and 28 February 1991, 846 Squadron flew over 1,000 hours undertaking 500 sorties which averaged 80 per week. These included assisting the USS *Tripoli* and battleships USS *Missouri* and *Wisconsin*. One of the biggest threats to the battleships, as they pounded Iraqi positions were from free floating mines known as 'Daisies' or from shore based 'Exocet' and 'Silkworm' missiles.

Both USS *Tripoli* and USS *Princetown* were hit by mines and 846 Squadron Sea Kings helped to lift the injured back to *Argus* for treatment. A priority task for the Sea Kings during this phase of the war was to act as mine spotters flying ahead of *Argus*.

During these searches the squadron found two mines. Once detected their position was marked and the Sea King remained on station until the arrival of an EOD team who were based on USS *Tripoli* or USS *Missouri*. 846 Squadron made good use of their NVG skills and the accuracy of the NAVSTAR/TANS2 system during a night search and the successful recovery of a USMC F-18 pilot,

Top left: An 846 Squadron Sea King (ZF118/VL) and HMS *London's* Lynx aboard RFA *Argus* as she steams along in the Gulf. *(Patrick Allen)*

Left: 846 Squadron Sea King (ZD476/VJ) flying ahead of RFA *Argus* in the Gulf in January 1991. The fresh pink paint shows where the Union Flag, painted on during their transit down to the Gulf, had now been removed once the ship was on station and prior to the squadron going ashore to undertake desert training exercises. *(Patrick Allen)*

who once located, was picked up by a USN Sea Hawk. During the securing of Kuwait City three of the squadron's Sea Kings were assigned to Special Forces duties and led the helicopter flight used to insert SAS/SBS troops onto the roof of the British Embassy in Kuwait. The squadron operated around the Kuwait City area for over 10 days undertaking numerous tasks which included transporting the Prime Minister and the Defence Secretary during their visits. In the six weeks of 'Operation Desert Storm' 846 Squadron Sea Kings flew over 1,200 hours on land and sea.

'A' and 'B' FLIGHT–HELICOPTER DELIVERY SERVICE

'A' Flight, 846 Naval Air Squadron started to operate the HDS Flight on 22 August 1990 aboard the 23,384-tonne RFA *Fort Grange*. They were replaced by 'B' Flight on 6 December 1990 with the 'A' Flight personnel returning to RNAS Yeovilton and joining 848 Squadron to deploy into the desert in January.

The squadron complement aboard RFA *Fort Grange* comprised two Sea King HC Mk4s (ZF118/ZF119) which stayed with RFA *Fort Grange* from August until February 1991, three full flying crews (nine) plus 21 support personnel, including helicopter maintainers etc. The HDS flight was crucial to the re-supply of the Royal Navy's warships operating in the Gulf. RFA *Fort Grange* was the supermarket for the fleet, and the helicopters the delivery vehicles. Stores, ammunition and missiles were ordered by ships, usually during the night. These orders were then split into packaged loads, weighed and sorted into a flight delivery sequence. Every item on the ship should already have been weight recorded and loads prepared for easy delivery.

Underslung loads for long range HDS transits are not recommended although some items of equipment such as Sea Dart ship-to-air missiles are often flown underslung. For the majority of the time stores are prepared for internal carriage. This allows faster transit speeds and gives the helicopter better manoeuvrability should it be attacked. The *Fort Grange* based Sea Kings had to perform

Top right: Two of 846 Squadron Sea Kings ZA296/VK and ZD478/VM land on RFA *Fort Grange* to take on fuel during a mission in the Gulf. 846 Squadron had Sea Kings aboard RFA *Argus* (4) and two (green) Sea Kings operating the helicopter delivery service (HDS) based on RFA *Fort Grange*. *(Patrick Allen)*

Right: Both 845 and 848 Squadron initially deployed to Al Jubayl in early January and undertook an intensive desert and NVG training programme prior to moving west into the desert to the King Khalid Military City (KKMC) area. On arrival at Al Jubayl aircraft were immediately fitted with 'Granby' enhancements including sand filters, GPS/TANS 2, defence aid suite including M130 chaff and flare dispensers. *(RNAS Yeovilton)*

a variety of different methods to transfer stores to ships including (VERTREP) Vertical Replenishment where underslung loads are carried across to a ship which is sailing alongside RFA *Fort Grange*. When delivering stores to small ships in rough conditions, or those without adequate deck space, or when high masts/aerials are a hazard, 'High-Line' transfers may be the only option. A line is attached to the winch cable allowing the helicopter to hover into the wind away from the ship. Cargo can then be sent down on the winch and pulled across to the ship using the 'High-Line'. This gives the helicopter freedom of movement away from the ship and reduces the need to make numerous approaches.

'B' Flight Sea Kings frequently undertook HDS sorties high into the Gulf off Kuwait which lasted six hours or more. On one occasion an HDS sortie lasted 9 hours 50 minutes without shutting down engines with the helicopter refuelling from ships along the way including a number of (HIFR) Helicopter Inflight Refuellings. With such long sea transits, often in high temperatures, the side cockpit windows were removed to help air circulation inside the cockpit during the hotter months. During these long range transits crews needed constant fluid replenishments, particularly during the hotter periods i.e. August–September. Small onboard refrigerators were installed in the helicopters and proved invaluable. During action stations HDS flights often had to be undertaken during the night requiring NVG deck landings. As one of the first Sea King Flights to operate in the Gulf they received some of the first 'Operation Granby' enhancements. These included ALQ 157 IR jammers, M130 chaff/flare dispensers and the 'Trimble' Satellite GPS system. This comprised a small self-contained unit which would display the aircraft's Lat/Long, still retaining the original Doppler TANS the GPS information needed to be manually transferred. With TANS-2 fitted both Doppler and GPS information is collated automatically with both systems crosschecking each other. During certain periods of the day/night the GPS can have difficulty in obtaining enough satellites for a fix and the Doppler system takes over until the GPS can update. The flying standards required for the HDS flights were high and extremely demanding, and must have been one of the most punishing for any supermarket delivery service. During the Gulf War period the two Sea Kings aboard RFA *Fort Grange* flew over 400 hours and from August 1990 until March 1991 they totalled over 1,000 flight hours.

DESERT KINGS — 845 and 848 NAVAL AIR SQUADRONS

By Christmas 1990 the entire front line strength of the Royal Navy's troop lifting Sea King force had been committed to 'Operation Granby'. As the crisis deepened, 845 Squadron began preparing their Sea King HC Mk4s for deployment into the desert to support the 1st (BR) Armoured Division in Saudi Arabia. On 6 December 1990, 848 Naval Air Squadron was reformed at RNAS Yeovilton with six Sea King HC Mk4s taken from reserves and 707 Naval Air Squadron. 848 Squadron had last been active in 1982 when it saw action in the Falklands operating the Wessex HU Mk5. The Commando Helicopter Operations and Support Cell (CHOSC) was also deployed. Their task was to integrate the RN Sea Kings into a combined RAF/RN Support Helicopter Force.

At 08.45, on Friday 21 December 1990 six 845 Naval Air Squadron Sea King HC Mk4s lifted from RNAS Yeovilton followed by six 848 Squadron Sea Kings bound for Southampton Docks and loading aboard the SS *Atlantic Conveyor II*. Arriving at Al Jubayl on 6 January 1991 the 12 Sea Kings were met by their 300 aircrew and maintainers. Along with RAF Puma and Chinook Support Helicopters they undertook an intensive training period as crews and aircraft became acclimatised to desert operations. This involved desert style tactical flying techniques and night vision goggle training. By the end of the short training programme all pilots and aircrew were capable of undertaking safe night operations, transitting at below 75 feet across the desert wearing AR5 NBC respirators and night vision goggles. Like the 846 Squadron Sea Kings all 845 and 848 Squadron aircraft received additional operational enhancements. These included sand filters, the new GPS-RNS 252 Super TANS-2, chaff and flare dispensers, etc.

On 22 January the Sea Kings left Al Jubayl and headed west for King Khalid Military City (KKMC). Here they undertook numerous exercises supporting 1st (BR) Armoured Division as newly arrived troops were brought

Left: 846 Naval Air Squadron Sea Kings on patrol in the Gulf. All 846 Squadron aircraft were armed with 7.62mm GPMG door guns. *(Patrick Allen)*

up to battle readiness. This training included a number of rehearsals at breaking through Iraqi style positions and defensive obstacle zones. The primary mission of the helicopters assigned to the Middle East Support Helicopter Force was casualty evacuation. The Sea Kings spent many hours on casevac exercises as medics and helicopter crews perfected their roles. Additional medics were carried as part of the helicopter's crew and stretcher bearers, doctors and medical orderlies all needed to practice their individual skills, particularly at night or under simulated NBC warfare conditions. During the build up phase helicopters also undertook numerous real casevac missions taking the injured, many from road traffic accidents, and the sick to Field Hospitals for treatment. A Royal Navy Sea King pilot Lt Commander Peter Nelson received the Air Force Medal when he flew 40 miles behind Iraqi lines, at night, in driving rain and visibility less than one mile to rescue an injured soldier and fly him back to an Army Field Hospital.

Just before the land battle on 21 February 1991 two spare Sea King HC Mk4s (ZE425/ZA291) were flown to the Gulf from RNAS Yeovilton by USAF C-5A Galaxy transport aircraft. This deployment brought to 520 the total of Yeovilton personnel in the Gulf. These two Sea Kings were eventually assigned to 846 Squadron with ZE425 returning home aboard RFA *Argus* a month later and ZA291/VN deploying aboard RFA *Fort Grange* to Bangladesh on 'Operation Mana'. Both these Sea Kings (ZE425/ZA291) were the 'Granby' fleet leaders and fitted with all the enhancements including IR Jammers and Missile Approach Warners (MAWS) etc.

As the 1st (BR) Armoured Division moved westward, so the Support Helicopter Force followed. By 'G' Day, Sunday 24 February they were all prepositioned at their respective Forward Operating Bases instantly available for tasking as the land battle began.

Allied casualties remained low, although the Support Helicopter Force was kept busy moving injured Iraqi troops and thousands of enemy prisoners of war (EPWs). During the 100-hour war the Sea Kings of 845 and 848 Squadrons continued to provide support to the Division as it advanced 135 miles into Iraq and Kuwait. During this period weather conditions were extremely poor with low cloud and heavy rain making flying difficult, particularly at night, wearing NVGs. After the advance and Kuwait City had been liberated the Sea Kings remained busy transporting essential personnel which included Explosive Ordnance and Demolition (EOD) teams who were needed to make the city and battlefield safe. Both 845 and 848 Squadron Sea Kings flew over four times their normal flight hours during the air/land war phase and operating from harsh desert forward operating bases Sea King readiness remained high.

Above: 21 December 1990 and 845 and 848 Squadron Sea Kings arrive at Southampton docks bound for the Gulf aboard MV *Atlantic Conveyor*. 845 Squadron Sea Kings are easily recognisable by having a small red mark in one of their cabin windows. A 707 Squadron Sea King (green) ZA292/ZW arrives in support of the 12 aircraft and to collect the 845/848 aircrew. *(RNAS Yeovilton)*

With 845 Naval Air Squadron remaining in Kuwait with three aircraft (ZD477/A, ZA312/B, ZA313/E) both 846 and 848 Squadron Sea Kings embarked aboard RFA *Argus* and sailed for home arriving off Portland on Tuesday 2 April 1991. The same day, 848 Squadron Sea Kings still retaining their white invasion stripes, flew to RNAS Portland where they regrouped for a formation arrival at RNAS Yeovilton at 14.00 hours. 846 Squadron remained on board *Argus* until she arrived off Spithead the following morning. After being received by the C-in-C Fleet Admiral Sir Jock Slatter KCB, LVO, 846 Squadron departed *Argus* on the morning of 3 April and regrouped at Fleetlands for a formation arrival at RNAS Yeovilton at 14.00 hours.

'OPERATION GRANBY'
ROYAL NAVY SEA KING HC MK4s

846 NAVAL AIR SQUADRON — RFA *ARGUS*, 'C' and 'D' FLIGHTS 4 a/c
ZA293/VP, ZA296/VK, ZD476/VJ, ZD478/VM

846 NAVAL AIR SQUADRON — RFA *FORT GRANGE*, 'A' and 'B' FLIGHTS 2 a/c
ZF118/VL and ZF119/VO

845 NAVAL AIR SQUADRON SUPPORT HELICOPTER FORCE MIDDLE EAST
ZD477/A, ZA312/B, ZD480/C, ZF117/D, ZA313/E, ZG820/F

848 NAVAL AIR SQUADRON SUPPORT HELICOPTER FORCE MIDDLE EAST
ZA298/WA then WJ, ZA314/WM. ZE427/WB then WK, ZE428/WC then WL, ZG821/WE then WN, ZG822/WF then WO

FLOWN INTO THEATRE ON 21/2/91 USAF C-5A GALAXY-70/0460/436MAW
Sea King ZE425 and Sea King ZA291

Note: At least three of 846 Naval Air Squadron Sea Kings ZA291/ZA293/ZA296 flew with the squadron during the Falklands Conflict. ZA293/ZA296 had totalled around 4,000 airframe hours by the start of 'Operation Granby'.

'C' FLIGHT 826 NAVAL AIR SQUADRON 'OPERATION GRANBY'

'C' Flight, 826 Naval Air Squadron arrived in the Persian Gulf on 13 December 1990 relieving 'D' Flight, 826 NAS who had been operating in the region since August 1990 based for a period of that time aboard the Dutch ship HNLMS *Zuiderkruis*.

826 Naval Air Squadron is one of the four front line Royal Navy ASW squadrons and is home based at RNAS Culdrose. Formed in 1983 the squadron comprises of four Flights A to D operating the Sea King HAS Mk5 primarily in the ASW role deployed aboard Royal Fleet Auxiliary ships or frigates. The flexibility of the Sea King and the self-contained nature of the squadron's Flights made them ideal for both the detached duties and variety of tasks undertaken during the Gulf conflict.

Initially operating from HNLMS *Zuiderkruis*, 'C' Flight was primarily tasked with enforcing United Nations sanctions by patrolling areas in the southern Persian Gulf, inspecting and interrogating any previously unidentified surface units. In addition they were also tasked with long-range helicopter stores delivery (HSD) to the many warships deployed throughout the region.

By early January 1990 with the worsening political situation with Iraq the 'C' Flight Sea Kings were tasked to assist the Multinational Mine Countermeasures Force (MCMVs) in forming a rapid reaction, mine detection and destruction unit. This comprised the Sea King HAS Mk5 using its airborne visual sighting equipment plus an onboard ordnance disposal team from the Royal Navy Fleet Diving Units ready for immediate deployment.

During their Gulf deployment the ASW sonar equipment was removed from the two Sea King HAS Mk5s allowing extra cabin space and an increase in operating weights. To undertake the surface search and mine countermeasures roles the Sea King HAS Mk5's equipment now included:

Standard Sea Searcher radar, GPS Trimble Navigation Equipment, ALQ 157 Infra Red Jammers, M-130 Chaff and Flare Dispensers, 7.62mm GPMG Door Gun, SAR Equipment, Plain and Secure Speech Radios, Radar Warning Receivers (RWR) plus EOD Explosives and Diving Equipment. Crews comprised: two Pilots, one Observer, one Aircrewman, one Diving Supervisor and three Divers.

Additional role equipment included: 'Menagerie' (ECM) equipment, 'Sandpiper' Forward Looking Infra Red (FLIR) and hand held thermal imaging systems and the 'Demon' camera mine hunting video system.

At the start of the war on 17 January 1991, 'C' Flight was operating alongside Allied warships in the protection of supply and auxiliary shipping in the Gulf region. With the imminent onset of a ground offensive 'C' Flight was split on 26 January with one Sea King deploying to RFA *Argus* and one to RFA *Sir Galahad*. The Landing Ship Logistics (LSL) RFA *Sir Galahad* was used as the forward operating base in support of MCMV operations in the region. Operating from the RFA *Sir Galahad* 'C' Flight was initially tasked in support of the MCMV forces in preparing 'safe areas' for Allied warships to carry out Naval Gunfire Support missions and for a possible amphibious assault by the USMC.

With the majority of support ships situated in the Dorra Oilfield area, USS *Tripoli*, RFA *Sir Galahad* plus Royal Navy Mine Countermeasures Vessels (MCMVs) and their escorts were positioned 30nm off the Kuwaiti coast to begin their task of clearing channels into the selected areas or 'boxes' required for the planned amphibious assault. This task included an 'initial look' sortie by a 'C' Flight Sea King along the coastline to within 8nm of Kuwait City, under an airborne Combat Air Patrol (CAP) and armed escort. During this early period of the war the increased threat of attack by Exocet and Silkworm missiles forced Allied ships to deploy further south which increased the Sea Kings' transit times during these surface search missions. With longer transit and sortie times the Sea Kings made maximum use of all available ships to 'hot' refuel or HIFR from. These included the USS *Tripoli*, USS *Missouri* plus RN Type 22s including HMS *London* and *Gloucester* etc. The Sea King, quick reaction, airborne mine clearance unit was put to good use on 18 February 1991 when both the USS *Tripoli* and USS *Princetown* both struck mines. The 'C' Flight Sea King was one of the first aircraft on task clearing a route and

Below: 846 Squadron Sea King HC Mk4 (ZA296) and a Royal Navy Lynx seen in front of a burning oilwell in Kuwait, February 1991.
(DPR Royal Navy)

assisting the safe transit of the surrounding task force in the area.

After the start of the ground offensive on 24 February and the cancellation of the proposed amphibious assault, 826 Squadron Sea Kings concentrated on their surface search role patrolling the sea areas around Kuwait City and Faylaka Island. The Sea King Mk5's superior avionics and radar fit made them an ideal platform to monitor the coastline and to control RN Lynx and USMC AH-1J Cobra helicopters as they searched for targets in the region. It was during one of these patrols that a 'C' Flight Sea King became the first Royal Navy helicopter to land in Kuwait City on the morning of 27 February 1991. During the following day the Flight returned to deploy survey teams into the Kuwait Harbour area.

With the majority of EOD taskings completed 'C' Flight airlifted both the Fleet Diving units ashore to Mina Al Shuaba on 5–6 March to carry out harbour clearance duties and they continued to assist in the Kuwait area until 16 March when the Flight embarked aboard the RFA *Fort Grange*. The Flight continued to undertake mine search and long-range HDS missions until they were relieved by 'D' Flight on 27 April 1991. Within a few weeks RFA *Fort Grange* along with the two Sea King Mk5s from 'D' Flight 826 Naval Air Squadron plus a Sea King HC Mk4 from 845 and 846 Naval Air Squadrons were then deployed to Bangladesh in support of 'Operation Mana'.

During 'Operation Granby' the two Sea King HAS Mk5s from 'C' Flight 826 Naval Air Squadron flew a total of 117 missions locating around 30 per cent of all moored mines found during the conflict. Their airborne mine clearance (EOD) team destroyed 17 of these mines becoming the single most effective airborne mine countermeasures unit operating in the Persian Gulf. Between 12 December 1990 and 27 April 1991 the two Sea Kings of 'C' Flight, 826 NAS flew 403 missions, a total of 724 flying hours locating 44 moored and floating mines. During their 106 days at sea, each crew averaged 241 flight hours on 135 missions. During the entire deployment the only role not undertaken by the flight was their primary role of ASW showing the versatility and adaptability of both the Sea King and Royal Navy aviators.

826 NAVAL AIR SQUADRON 'OPERATION GRANBY' and 'OPERATION MANA'

'D' Flight, 826 NAS from 22 August 1990 to 12 December 1990.
(Sea King HAS Mk5s: ZE422 and XV661 'Sandpiper fit')

'C' Flight, 826 NAS from 12 December 1990 to 27 April 1991
(Sea King HAS Mk5s: ZA137 and XZ575)

'D' Flight, 826 NAS from 27 April 1991 to June 1991 (Op. Mana)
(Sea King HAS Mk5s: ZA137 and XZ575)

Note: Sea King HAS Mk5 XZ575/138 was lost in an accident during 'Operation Mana' when a main rotor blade struck the deck netting of RFA *Fort Grange*. The Sea King was able to make a controlled ditching into the sea and crew and passengers safely vacated the aircraft which subsequently sank. There was no injury or loss of life.

Below: 826 Squadron Sea Kings operated from RFA and Dutch ships during the Gulf War. Two of their Sea Kings deployed to the Gulf in August 1990 and the two pictured here deployed just post war. 826 Squadron Sea Kings were modified to undertake the surface search role and proved vital in the mine clearance role. Modifications included the removal of their ASW equipment and the fitting of special video search cameras, M-130 Chaff and Flare dispensers, Trimble GPS satellite navigation system, ALQ 157 IR jammers, Yellow Veil ECM, and Sandpiper Forward Looking Infrared (FLIR). These Sea Kings operated from several different ships including two months on the Dutch tanker *Zuiderkruis* and finally aboard the RFA *Argus* and RFA *Sir Galahad*, searching for and clearing mines from the Kuwaiti ports. 826 Squadron Sea Kings later deployed to Bangladesh for 'Operation Mana' aboard RFA *Fort Grange*. *(RNAS Culdrose)*

SECTION 23
'OPERATION HAVEN'

In early April 1991 a massive Allied effort was launched to provide humanitarian assistance to help thousands of Kurdish refugees stranded in the mountains between Turkey and Iraq. The area was inhospitable, difficult to reach, and the only way to deliver urgently-needed supplies was by air. This was initially undertaken by C-130 Hercules transport aircraft until they were joined by a huge Allied helicopter force. Called 'Operation Provide Comfort' this was later changed to 'Operation Haven' when it was decided that Allied troops would secure a 'Safe Haven' in Northern Iraq for the Kurdish refugees. Allied military helicopters were essential for the success of both these operations.

3 Commando Brigade Royal Marines were assigned the task of deploying to Kurdistan and would take their own support and attack helicopter force. As mountain warfare specialists the Royal Marines were best suited for operations both in the mountain areas and in helping to secure the proposed 'safe haven' area in Northern Iraq. They could also boast their own Medical Squadron to assist the returning Kurdish families. To support 3 Commando Brigade Royal Marines, 846 Naval Air Squadron, who only just having returned from the Gulf on 3 April, were recalled from leave.

By mid-April most of 846 and 848 Naval Air Squadron Sea Kings were undergoing deep maintenance at RNAS Yeovilton. Several of the helicopters had suffered 290 air-frame cracks as a result of their Gulf operations and all needed inspection. A team from the Royal Navy's Mobile Aircraft Support Unit (MASU) were deployed to Yeovilton and worked around the clock repairing the aircraft.

Within a week 846 Naval Air Squadron had planned and prepared their deployment for 'Operation Haven'. Using all available Sea Kings, 846 Squadron was increased in size to form a 12-aircraft squadron. Aircraft were taken from both 848 and 845 Squadrons back at Yeovilton, and from diverting aircraft returning from the Gulf by sea. The Sea Kings that had returned from the Gulf had their desert camouflage washed off although there was not enough time to remove their white invasion stripes, painted on prior to 'Desert Storm'. Three of 845 Squadron Sea Kings returning from the Gulf aboard the SS *Baltic Eagle* were diverted to Turkey and these aircraft retained their desert camouflage.

On Sunday 21 April 1991 eight Sea King HC Mk4s departed RNAS Yeovilton to join RFA *Argus* in Portsmouth. Flown by 848 Squadron crews, 845 and 846 Squadron personnel would fly ahead to meet the ship as she arrived in Turkey. During the sea transit 848 Squadron maintainers, plus three aircrew would look after the aircraft and complete any outstanding maintenance and necessary check flights. Once completed both 845 and 848 Squadron aircraft would receive their new 846 Squadron codes.

On the evening of Sunday 21 April 1991, within two and a half weeks of arriving back from the Gulf, RFA *Argus* sailed from Portsmouth carrying eight Sea King helicopters, six Lynx and four Gazelles from 3 Commando Brigade Air Squadron, plus 50 assorted vehicles.

On Monday 29 April Lt Rob Saddington RN and Lt Dave Kistrick RN, undertook the last 848 Naval Air Squadron flight when they departed RFA *Argus* in Sea King ZA298/WJ for an HDS sortie to RAF Akrotiri in Cyprus. The following morning, Tuesday 30 April, 846 Squadron personnel were flown aboard RFA *Argus* as she steamed past Karatas Point off the coast of Turkey to fly their eight Sea Kings to Incirlik airbase.

Regrouping at Incirlik, the squadron later that day flew east across Turkey, to Silopi on the Turkey/Iraq border. During the 400nm transit the Sea Kings refuelled at Diyarbakir before continuing to Silopi arriving in the early evening. Deploying with the minimum of personnel, the main support elements of the squadron would follow in road transport taking several days to cross Turkey. The Sea Kings by early the next morning were available for tasking.

Below: Last flight by an 848 Squadron Sea King, ZA298/WJ recoded 'VP', on 29 April 1991 during an HDS sortie to Cyprus as RFA *Argus* steams towards Turkey loaded with Sea King, Lynx and Gazelle helicopters. *(Patrick Allen)*

SILOPI

Most of the Kurdish refugees had made their camps along the Turkey/Iraq/Iran mountain borders at altitudes in excess of 8,000 feet. The nearest military airfield was at Diyarbakir about 175 nm to the west and this had become the major airhead for both incoming supplies and for many of the Allied helicopters. Silopi, which is a small town situated close to the Iraqi border and the mountains, was ideally suited for the Allied helicopters assigned to securing the Safe Haven. By early April a large field outside the town had been purchased by the Allies and used as a forward operating base for the helicopters operating out of Diyarbakir.

The field was ideally suited for helicopters and for those involved with the refugees and 'Operation Haven'. Silopi became the main operating base for 846 Squadron although they periodically forward deployed aircraft into Iraq. By late April the field at Silopi saw more aircraft movements in a single day than London Heathrow airport. During April, Silopi became home to almost every type of helicopter in the coalition force inventory including those from the US Army, Navy and USMC. The British (FARP) forward arming and refuelling point was handling over 70 refuels and using over 60,000 gallons of AVTUR a day. The US FARP was also breaking records as helicopters lined up at their five refuel points.

'OPERATION HAVEN'

US Special Forces and M&AW Cadre troops from the Royal Marines had been deployed into the mountains to provide much needed medical assistance and to operate helicopter landing sites. These troops proved essential in the organising and distribution of incoming supplies and building up trust with the Kurdish families.

Having undertaken immediate life-saving missions, the Allies' priority now turned to

Below right: Sea King ZE425/G waits, rotors running, as Kurdish refugees are quickly loaded aboard. The mountain landing site at Sinat was located above one of the mountain valleys which had recently been gassed by Saddam Hussein. *(Patrick Allen)*

Below left: 846 Squadron Sea King ZE425/G arrives at a mountain refugee site to collect Kurdish families and return them to Zakhu on 2 May 1991. The landing site was run by troops from the RM M&AW Cadre and US Army 10th Special Forces. Over 250 Kurdish refugees were flown out by 846 Squadron in a single afternoon. *(Patrick Allen)*

getting the Kurdish refugees off the mountains and back to their homes in Northern Iraq. It was decided that Allied troops would be deployed into Northern Iraq, along the Silopi valley bowl to form a Safe Haven of approximately 100–70kms. This would enable the Kurdish refugees to leave their mountain camps and to return home with some form of Allied protection. Initially this protection would be provided by Allied troops until handed over to the United Nations. With recent outbreaks of typhoid and cholera in a number of mountain camps and with the onset of summer, time was extremely short. The Kurds had been relying on melting mountain snow for their water source. With rising temperatures the snow line was rapidly melting and would soon be gone. Helicopters would be unable to keep up with the increased demand water supplies would create.

3 Commando Brigade Royal Marines, plus the US Army's 3-325th Infantry, US Marines and French troops were assigned the task of forming the Haven which was split into two groups called Task Force Alpha and Task Force Bravo:

Task Force Alpha comprising the US 10th Special Forces Group together with 40 Commando Royal Marines, would continue to work in the mountains helping prepare the Kurdish families for their return home.

Task Force Bravo comprising 45 Commando Royal Marines, Dutch Marines from 1ACG plus US and French troops would form the Safe Haven area and secure the region for the Kurds' return.

'Operation Haven' was not a combat operation although the potential for renewed hostilities with the Iraqi military was high. The priority of the mission was to save the lives of Kurdish families in the mountains, and then to return them to their homes in Northern Iraq. During all phases of the operation US Commanders met daily with their Iraqi counterparts. Each phase of the operation was explained in detail and any Iraqi military or civilian grievance was sorted out. This also allowed the Iraqi military the opportunity to disarm or remove their forces from an area prior to the arrival of the Allied troops.

KURDISH REFUGEES

During their first afternoon in theatre two 846 Squadron Sea Kings (ZG821/VL and ZE425/G) moved over 250 Kurdish refugees from their mountain camp at Sinat, back to their homes in Zachu, Northern Iraq. Sinat Camp was located high above one of the mountain valleys which had recently been gassed by Saddam Hussein.

Early the following day on 2 May the squadron was tasked to support 3 Commando Brigade as they began to secure the Safe Haven area. This included a number of helicopter deployments and one which landed up inside Saddam Hussein's Summer Palace and a short exchange of words with the Palace Guards. During the first day other objectives included a strategically placed bridge on the road east to Bafuta and an Iraqi airfield at Sirsenk. Bombed by USAF B-52s during 'Desert Storm' the airfield when repaired, would provide a valuable air head for the Allied troops.

Bottom: 846 Squadron's tented operating base at Silopi, Turkey. The site soon turned into Tent City as Silopi became home to other Allied helicopter units. *(Patrick Allen)*

Below: 846 Squadron Sea Kings join RAF Chinooks at the refuelling site at Silopi during the first days of 'Operation Haven'. *(Patrick Allen)*

Bottom: RM troops board 846 Squadron Sea Kings (ZD477/A-ZA314/VK) in Iraq to be moved forward to secure a new location in northern Iraq during the first day of 'Operation Haven'. *(Patrick Allen)*

Below: 846 Squadron Sea King (ZA314/VK) collects RM troops in Iraq during the early phase of 'Operation Haven'. *(Patrick Allen)*

A further wave of Sea Kings was tasked with inserting troops into the mountain village of Amadiya and throughout the region. Over the following days the Sea Kings were kept busy re-deploying and re-supplying troops as they began their diplomatic task of winning 'hearts and minds' and persuading the Kurdish families still in the mountains that it was now safe to return.

One of the major hazards for both men and helicopters operating in the region was from mines. These had been liberally scattered throughout the area and there was the possibility of booby traps having been left by Iraqi troops for either the Allied troops or returning Kurds. Helicopter crews needed to check under their helicopters for mines, particularly on landing. Dead animals in an area was always a good indication of their presence.

As the Allied troops moved further into Iraq the potential for an outbreak of hostilities with Iraqi forces remained high. This was evident by the short and violent firefight which took place between the Palace Guard and Royal Marine Commandos outside one of the Summer Palaces. Although disorganised the Iraqis had the potential to muster over 40,000 troops from their 16th, 44th and 88th Divisions based in the north plus elements of the Palace Guard.

During 'Operation Haven' Allied helicopters were continually monitored by AWAC aircraft who provided an emergency and flight following service. At night, this 'eye in the sky' role was undertaken by USN Hawkeyes. Continual CAPs were flown by F-15s, F-16s with close air support provided by the A-10s who patrolled the mountain areas.

Attack helicopter escorts if required were provided by TOW-armed Lynx Mk7s from 3 Commando Brigade Air Squadron, TOW and Sidewinder-armed USMC AH-1Ws and US Army AH-64s. The AH-64s from 6/6th Cavalry Squadron fitted with a mix of Hellfire missiles, 2.75in rockets, 30mm cannon and long range fuel tanks undertook regular day and night patrols throughout the region. Using their Pilots' Night Vision System (PNVS) and TADS they could keep watch for clandestine Iraqi activity in the region.

HOT, HIGH and HEAVY MOUNTAIN FLYING

846 Squadron's previous mountain flying training proved invaluable in Kurdistan. To help compensate for the decreased air density due to both altitude and temperature, the Sea Kings restricted their weight to eight combat troops plus 3,000lb of fuel which would leave them a reasonable power margin for mountain operations. Working at the limits of their performance Sea Kings would be burning around 1,200lb of fuel per hour.

Most helicopters, including the Sea King, are designed for optimum performance at low levels. As air density decreases either by higher temperatures or altitude, or both, more power is required along with alterations in rotor pitch to compensate for the thinning air. Up to certain altitudes, both power and rotor pitch settings can increase efficiency, after this, both power and rotor efficiency fall and the two effects become cumulative. This can be critical during low speed manoeuvres, landing or hovering out of ground effect when high power margins are required.

Prior to each mission, pilots needed to check the appropriate operator's manual and look up the helicopter's density altitude and performance graphs and to check their aircraft's engine power performance (PPI) from the Form 700 and to check the aircraft's weight. With 'Granby' enhancements some aircraft weighed more than others. For example a fully modified aircraft such as ZE425/G weighed in at 14,400lb.

These types of hot, high and heavy mountain flying conditions can result in blade stall or reduced aircraft stability. In mountainous terrain winds can be highly unpredictable, with their associated up and down-draughts, mountain wave turbulence and these can vary in intensity without warning. In the mountains pilots needed to recce every landing site, check their power margins and pick good escape and take-off routes. This type of flying is always demanding both on the aircrew and helicopters.

By early May the number of major refugee camps had been steadily reduced as the success of 'Operation Haven' grew. With the help of Allied troops and helicopters Kurdish families were coming away from the mountains in ever increasing numbers. One of the regular missions undertaken by the 846 Squadron Sea Kings was to fly UN and Kurdish leaders to the different refugee camps in an attempt to persuade the more reluctant families to return home. By late May the first UN officials had arrived to take over the running of the 'Safe Haven' and by late June 846 Squadron had returned to Yeovilton. During the deployment lasting from 29 April until 24 June, 846 Squadron Sea Kings flew over 1,200 hours without major incident, a credit to the squadron maintainers. This was even more remarkable considering the majority of the Sea Kings had previously been deployed in the desert with 845 and 848 Squadrons during operation 'Desert Storm'.

'OPERATION MANA'

As 'Operation Haven' got underway in Kurdistan, US and Royal Navy helicopters which included two Sea King HC Mk4s from 845 and 846 Squadron (ZG820/F and ZA291/VN) and two Sea King HAS Mk5s (XZ577/138 and 137) from 'D' Flight 826 Naval Air Squadron deployed to Bangladesh leaving the Gulf on 10 May 1991 aboard the RFA *Fort Grange* for 'Operation Mana'. Arriving off Bangladesh on 18 May the Sea Kings were tasked in support of the Bangladesh Cyclone relief operation and delivered more than 400 tons of stores within a two-week period, mainly to the small villages located on the three islands which had received the brunt of the storm. During the deployment the Sea Kings operated in some extreme weather conditions with one of the Commando Sea Kings being struck by lightning during a tropical storm, damaging one of its composite rotor blades and one experiencing an engine flame-out due to excessive water build up in its sand filters due to rain. On 1 June 1991 while embarked on RFA *Fort Grange*, 826 Squadron's, Sea King HAS Mk5 (XZ577/138) ditched and was written off, fortunately without injury to the crew.

Below: A pair of 846 Squadron Sea Kings (ZA314/VK-ZA298/VO) filled with Royal Marines bank round to begin the first assault on the Iraqi airfield at Sirsenk — code name 'Objective Lion'. *(Patrick Allen)*

'OPERATION HAVEN'
846 NAVAL AIR SQUADRON
SEA KING HC MK4s

ZA298 ex-848 Squadron coded 'WJ' then coded 846/VP.

ZA314 ex-848 Squadron coded 'WM' then coded 846/VK.

ZE425 one of two spares flown to the Gulf by USAF C-5A Galaxy and returned to RNAS Yeovilton on RFA *Argus* with 846 Sqn and redeployed for 'Operation Haven' coded 'G'.

ZE426 ex-848/845 Squadron coded 'A' recoded 846/VO.

ZE427 ex-848 Squadron coded 'WK' then coded 846/VN.

ZE428 ex-848 Squadron coded 'WL' then coded 846/VJ.

ZG821 ex-848 Squadron coded 'WN' then coded 846/VL.

ZG822 ex-848 Squadron coded 'WO' then coded 846/VM.

All the above aircraft deployed to Turkey aboard RFA *Argus*.

Sea King ZD477 ex-845 Squadron coded ZD477/A.

Sea King ZA312 ex-845 Squadron coded ZA312/B.

Sea King ZA313 ex-845 Squadron coded ZA313/E (remained at Iskenderun u/s).

The above three aircraft arrived in Turkey via the MV *Baltic Eagle* diverted from their return to the UK from Saudi Arabia.

Sea Kings ZG821 and ZG822 were the last two Royal Navy Sea King HC Mk4s to be built by Westland and were delivered to the RN in late 1990. Joining 848 Squadron with just their test flight and delivery hours both aircraft had by the time they deployed to Turkey flown: ZG821/WN/VL 173 hours and ZG822/WO/VM 243 hours.

'OPERATION MANA'

Four Royal Navy Sea Kings embarked aboard the RFA *Fort Grange* to provide cyclone relief aid in Bangladesh. The two month 'Operation Mana' finished in June 1991. Aircraft support was provided by two Sea King HAS Mk5s from 'D' Flight 826 Squadron plus one Sea King HC Mk4 from 845 Naval Air Squadron (ZG820) and one from 846 Squadron 'B' Flight (ZA291). 826 Squadron lost Sea King HAS Mk5 (XZ577/138) when it was written-off in theatre on 1 June 1991.

Below: 846 Squadron Sea King (ZA428/VJ), one of two Sea Kings undertaking the first insertion of RM troops onto the town football pitch of the Iraqi mountain town of Al Amadiyah on 3 May 1991. *(Patrick Allen)*

PRODUCTION LIST

WESTLAND SEA KINGS — ALL MARKS

C/N	Bld No.	Serial	Mark	F/F	D/D	Remarks
WA630	1	XV642	HAS1	7/5/69		to HAS2A
WA631	2	XV643	HAS1	10/6/69	29/1/73	to HAS 6
WA632	3	XV644	HAS1	7/8/69	19/8/69	to A2664
WA633	4	XV645	HAS1	9/7/69	11/8/69	w/off 13/1/72
WA634	5	XV646	HAS1	30/7/69	19/8/69	w/off 25/10/77
WA635	6	XV647	HAS1	6/9/69	26/9/69	to SAR Mk5
WA636	7	XV648	HAS1	14//9/69	6/10/69	to HAS 6
WA637	8	XV649	HAS1	21/9/69	6/10/69	to AEW2A
WA638	9	XV650	HAS1	2/10/69	6/11/69	to AEW2A
WA639	10	XV651	HAS1	17/10/69	7/11/74	to HAS 6
WA640	11	XV652	HAS1	12/11/69	3/12/69	to HAS 5 w/off 3/2/88
WA641	12	XV653	HAS1	24//11/69	9/12/69	to HAS 6
WA642	13	XV654	HAS1	9/12/69	5//1/70	to HAS 6
WA643	14	XV655	HAS1	24/12/69	20/1/70	to HAS 6
WA644	15	XV656	HAS1	13/1/70	4/2/70	to AEW2A
WA645	16	XV657	HAS1	21/1/70	11/2/70	to HAS 6
WA646	17	XV658	HAS1	3/2/70	26/2/70	w/off 3/2/83
WA647	18	XV659	HAS1	13/2/70	3/3/70	to HAS 6
WA648	19	XV660	HAS1	27/2/70	26/3/70	to HAS 6
WA649	20	XV661	HAS1	9/3/70	2/4/70	to HARNK5
WA650	21	XV662	HAS1	11/3/70	2/4/70	w/off 10/4/72
WA651	22	XV663	HAS1	15/4/70	22/5/70	to HAS 6
WA652	23	XV664	HAS1	6/5/70	1/6/70	to AEW2A
WA653	24	XV665	HAS1	17/6/70	2/7/70	to HAS 6
WA654	25	XV666	HAS1	23/6/70	2/7/70	to SAR Mk5
WA655	26	XV667	HAS1	14/7/70	6/8/70	w/off 12/12/74
WA656	27	XV668	HAS1	22/7/70	20/8/70	w/off 24/2/87
WA657	28	XV669	HAS1	12/8/70	3/9/70	to A2659
WA658	29	XV670	HAS1	18/8/70	3/9/70	to HAS 6
WA659	30	XV671	HAS1	26/8/70	1/10/70	to AEW2A
WA660	31	XV672	HAS1	17/9/70	5/10/70	to AEW2A
WA661	32	XV673	HAS1	18/9/70	1/10/70	to HAS 6
WA662	33	XV674	HAS1	9/10/70	9/11/70	to HAS 6
WA663	36	XV675	HAS1	27/11/70	11/1/71	to HAS 6
WA664	37	XV676	HAS1	27/11/70	16/12/70	to HAS 6
WA665	38	XV677	HAS1	18/12/70	22/1/71	to HAS 6
WA666	39	XV695	HAS1	/1/71	5/2/71	w/off 17/11/75
WA667	40	XV696	HAS1	18/1/71	4/2/71	to HAS 6
WA668	41	XV697	HAS1	18/2/71	10/3/71	to AEW2A
WA669	44	XV698	HAS1	19/3/71	2/4/71	w/off 5/6/89
WA670	45	XV699	HAS1	1/4/71	29/4/71	w/off 5/6/89
WA671	46	XV700	HAS1	9/4/71	27/5/71	to HAS 6
WA672	47	XV701	HAS1	14/5/71	4/6/71	to HAS 6
WA673	48	XV702	HAS1	26/5/71	14/7/71	w/off 21/3/74
WA674	50	XV703	HAS1	14/6/71	3/8/71	to HAS 6
WA675	51	XV704	HAS1	20/7/71		HAS2A from AEW
WA676	53	XV705	HAS1	23/8/71	21/9/71	to SAR Mk5
WA677	54	XV706	HAS1	17/9/71	8/10/71	to HAS 6
WA678	55	XV707	HAS1	23/9/71	8/10/71	to AEW2A
WA679	57	XV708	HAS1	7/12/71	5/1/72	to HAS 6
WA680	59	XV709	HAS1	20/12/71	7/1/72	to HAS 6
WA681	61	XV710	HAS1	19/1/72	1/3/72	to HAS 6
WA682	63	XV711	HAS1	10/2/72	20/3/72	to HAS 6
WA683	64	VX712	HAS1	8/3/72	20/4/72	to HAS 6
WA684	65	XV713	HAS1	18/4/72	19/5/72	to HAS 6
WA685	67	XV714	HAS1	15/5/72	9/6/72	to AEW2A
WA733	34	IN501	Mk42	14/10/70	15/3/71	to India

C/N	Bld No.	Serial	Mark	F/F	D/D	Remarks
WA734	35	IN502	Mk42	20/10/70	22/3/71	to India
WA735	41	IN503	Mk42	20/4/71	8/10/71	w/off 2/2/81
WA736	43	IN504	Mk42	15/3/71	17/6/71	to India
WA737	49	IN505	Mk42	14/6/71	14/10/71	to India
WA738	52	IN506	Mk42	23/7/71	23/8/71	w/off 18/6/83
WA744	56	89+50	Mk41	6/3/72	18/4/73	to WGN
WA745	58	89+51	Mk41	26/6/72	7/5/76	to WGN
WA746	60	060	Mk43	19/5/72	16/12/72	to RNoAF
WA747	62	062	Mk43	21/6/72	16/12/72	to RNoAF
WA748	66	066	Mk43	30/6/72	10/8/73	to RNoAF
WA749	68	068	Mk43	18/7/73	15/6/73	w/off 10/11/86
WA750	69	069	Mk43	30/7/72	1/4/73	to RNoAF
WA751	70	070	Mk43	15/8/72	1/4/73	to RNoAF
WA752	71	071	Mk43	30/8/72	29/1/73	to RNoAF
WA753	72	072	Mk43	9/9/72	15/11/72	w/off 30/4/77
WA754	73	073	Mk43	21/9/72	15/11/72	to RNoAF
WA755	74	074	Mk43	30/9/72	15/11/72	to RNoAF
WA756	75	89+52	Mk41	7/6/73	14/11/74	to WGN
WA757	76	89+53	Mk41	14/6/73	29/11/74	to WGN
WA758	78	89+54	Mk41	5/7/73	27/3/75	to WGN
WA759	81	89+55	Mk41	31/8/73	17/10/75	to WGN
WA760	82	89+56	Mk41	6/9/73	14/11/74	to WGN
WA761	83	89+57	Mk41	10/9/73	24/4/75	to WGN
WA762	89	89+58	Mk41	8/11/73	25/2/74	w/off 11/1/82
WA763	90	89+59	Mk41	21/11/73	20/3/74	to WGN
WA764	91	89+60	Mk41	2/1/74	20/3/74	to WGN
WA765	92	89+61	Mk41	10/1/74		w/off B/D remains to WGN
WA766	93	89+62	Mk41	24/1/74	27/3/74	to WGN
WA767	95	89+63	Mk41	19/2/74	27/3/74	to WGN
WA768	96	89+64	Mk41	13/3/74	19/4/74	to WGN
WA769	97	89+65	Mk41	26/3/74	14/5/74	to WGN
WA770	102	89+66	Mk41	5/6/74	11/3/75	to WGN
WA771	103	89+67	Mk41	25/6/74	9/8/74	to WGN
WA772	104	89+68	Mk41	5/7/74	2/10/74	to WGN
WA773	105	89+69	Mk41	19/7/74	13/9/74	to WGN
WA774	106	89+70	Mk41	9/8/74	4/10/74	to WGN
WA775	107	89+71	Mk41	21/8/74	9/10/74	to WGN
WA776	77	IN507	Mk42	17/7/73	28/9/73	w/off 22/8/77
WA777	79	IN508	Mk42	7/8/73	8/10/73	w/off 17/1/86
WA778	80	IN509	Mk42	14/8/73	22/10/73	to India
WA779	94	IN510	Mk42	27/2/74	19/7/74	to India
WA780	99	IN511	Mk42	10/4/74	27/6/74	to India
WA781	101	IN512	Mk42	4/5/74	4/7/74	to India
WA782	84	262	Com1	12/9/73	29/1/74	to Egypt
WA783	85	263	Com1	13/9/73	13/2/74	to Egypt
WA784	86	264	Com1	21/9/73	25/2/74	to Egypt
WA785	87	265	Com1	26/9/73	7/6/74	to Egypt
WA786	88	261	Com1	30/9/73	15/1/74	to Egypt
WA787	98	N16-098	Mk50	30/6/74	21/3/75	w/off 24/5/79
WA788	100	N16-100	Mk50	6/8/74	8/4/75	to RAN
WA789	112	N16-112	Mk50	16/10/74	16/4/75	w/off 13/7/86
WA790	113	N16-113	Mk50	24/10/74	22/4/75	w/off 30/11/76
WA791	114	N16-114	Mk50	13/11/74	11/7/75	to RAN
WA792	117	N16-117	Mk50	13/12/74	27/1/75	to RAN
WA793	118	N16-118	Mk50	3/1/75	18/2/75	w/off 21/9/75
WA794	119	N16-119	Mk50	4/2/75	25/3/75	w/off 9/5/77
WA795	124	N16-124	Mk50	25/3/75	1/5/75	to RAN
WA796	125	N16-125	Mk50	8/5/75	18/7/75	to RAN
WA797	108	4510	Mk45	30/8/74	22/12/77	to Pakistan
WA798	109	4511	Mk45	10/9/74	8/3/77	to Pakistan
WA799	110	4512	Mk45	17/9/74	22/10/75	to Pakistan

WESTLAND SEA KINGS — ALL MARKS — continued

C/N	Bld No.	Serial	Mark	F/F	D/D	Remarks
WA800	111	4513	Mk45	30/9/74	29/3/77	w/off 8/2/86
WA801	115	4514	Mk45	27/11/74		to Pakistan
WA802	116	4515	Mk45	10/12/74	8/11/77	to Pakistan
WA803	121	721	Com2	16/1/75	21/2/75	to Egypt
WA804	123	722	Com2	5/2/75	7/3/75	w/off 9/7/75
WA805	126	723	Com2B	13/3/75	16/5/75	to Egypt
WA806	129	WA806	Com2	11/7/75	19/8/75	to Egypt
WA807	130	WA807	Com2B	25/7/75	/3/76	to Egypt
WA808	131	WA808	Com2	22/8/75	23/9/75	to Egypt
WA809	135	WA809	Com2	4/9/75	23/9/75	to Egypt
WA810	136	WA810	Com2	16/9/75	7/10/75	to Egypt
WA811	138	WA811	Com2	8/10/75	31/10/75	to Egypt
WA812	139	WA812	Com2	10/10/75	31/10/75	to Egypt
WA813	141	WA813	Com2	22/10/75	11/11/75	to Egypt
WA814	143	WA814	Com2	4/11/75	25/11/75	to Egypt
WA815	144	WA815	Com2	11/11/75	25/11/75	to Egypt
WA816	145	WA816	Com2	18/11/75	11/12/75	to Egypt
WA817	146	WA817	Com2	3/12/75	30/12/75	to Egypt
WA818	147	WA818	Com2	11/12/75	30/12/75	to Egypt
WA819	148	WA819	Com2	9/1/76	28/1/76	to Egypt
WA820	149	WA820	Com2	21/1/76	14/2/76	to Egypt
WA821	150	WA821	Com2	27/1/76	14/2/76	to Egypt
WA822	122	WA822	Mk47	11/7/75	29/8/75	to Egypt (CU)
WA823	127	WA823	Mk47	7/8/75	28/1/76	to Egypt
WA824	133	WA824	Mk47	4/9/75	30/9/75	to Egypt
WA825	134	WA825	Mk47	25/9/75	15/10/75	to Egypt
WA826	137	WA826	Mk47	23/10/75	14/1/76	to Egypt
WA827	140	WA827	Mk47	5/11/75	14/1/76	to Egypt
WA828	128	QA20	Com2A	9/8/75	10/10/75	to Qatar AF
WA829	132	VIP	Com2C	9/10/75	26/1/76	to Qatar AF
WA830	120	89+61	Mk41	18/4/75	23/7/75	to WGN (2nd A/c)
WA831	142	RS01	Mk48	19/12/75	18/6/76	to Belgium (CU)
WA832	153	RS02	Mk48	8/4/76	20/5/76	to Belgium (CU)
WA833	154	RS03	Mk48	12/4/76	28/5/76	to Belgium (CU)
WA834	155	RS04	Mk48	7/5/76	28/5/76	to Belgium (CU)
WA835	157	RS05	Mk48	7/6/76	14/7/76	to Belgium (CU)
WA836	151	QA22	Com2A	10/3/76	17/5/76	to Qatar AF
WA837	152	QA21	Com2A	16/3/76	5/4/76	to Qatar AF
WA838	156	XZ570	HAS2	18/6/76	19/7/76	to HAS 5
WA839	158	XZ571	HAS2	8/7/76	1/9/76	to HAS 6
WA840	159	XZ572	HAS2	20/7/76	9/9/76	w/off 14/1/80
WA841	160	XZ573	HAS2	5/8/76	13/9/76	w/off 18/5/82
WA842	161	XZ574	HAS2	2/9/76	6/10/76	to HAS 6
WA843	162	XZ575	HAS2	8/9/76	6/10/76	to HAS 6
WA844	163	XZ576	HAS2	17/9/76	27/10/76	to HAS 6
WA845	164	XZ577	HAS2	5/11/76	13/12/76	w/off 1/6/91
WA846	165	XZ578	HAS2	1/12/76	18/1/77	to HAS 6
WA847	170	XZ579	HAS2	18/3/77	30/5/77	to HAS 6
WA848	171	XZ580	HAS2	25/3/77	13/6/77	to HAS 6
WA849	172	XZ581	HAS2	20/3/76	26/7/77	to HAS 6
WA850	173	XZ582	HAS2	13/7/77	1/9/77	to HAS 5 w/off 27/10/89
WA851	174	XZ585	HAR3	6/9/77	26/1/78	to RAF
WA852	175	XZ586	HAR3	22/9/77	7/2/78	to RAF
WA853	176	XZ587	HAR3	17/10/77	12/12/77	to RAF
WA854	177	XZ588	HAR3	10/11/77	16/1/78	to RAF

WESTLAND SEA KINGS — ALL MARKS — continued

C/N	Bld No.	Serial	Mark	F/F	D/D	Remarks
WA855	178	XZ589	HAR3	2/12/77	10/3/78	to RAF
WA856	179	XZ590	HAR3	4/1/78	13/3/78	to RAF
WA857	180	XZ591	HAR3	12/2/78	11/4/78	to RAF
WA858	181	XZ592	HAR3	1/3/78	3/5/78	to RAF
WA859	182	XZ593	HAR3	30/3/78	17/5/78	Ditched 22/9/92 (CAT 4/5)
WA860	183	XZ594	HAR3	27/4/78	12/6/78	to RAF
WA861	184	XZ595	HAR3	27/4/78	16/6/78	to RAF
WA862	185	XZ596	HAR3	23/5/78	1/8/78	to RAF
WA863	187	XZ597	HAR3	22/6/78	25/8/78	to RAF
WA864	191	XZ598	HAR3	11/8/78	29/9/78	to RAF
WA865	193	XZ599	HAR3	1/12/78	2/2/79	to RAF
WA866	186	SU-BBJ	Com2E	1/9/78	9/12/80	to Egypt
WA867	188	SU-ARR	Com2E	28/9/78	3/4/79	to Egypt
WA868	190	SU-ART	Com2E	18/10/78	3/4/79	to Egypt
WA869	192	SU-ARP	Com2E	4/12/78	3/4/79	to Egypt
WA870	166	XZ870	HC4	built to stage 2	1/77	see ZA290
WA871	167	XZ871	HC4	built to stage 2	2/77	see ZA291
WA872	168	XZ872	HC4	built to stage 2	3/77	see ZA292
WA873	169	XZ873	HC4	built to stage 2	4/77	see ZA293
WA874	189		Mk43A	6/7/78	7/9/78	to RNoAF
WA875	194	XZ915	HAS2	7/1/79	30/3/79	w/off 6/3/81
WA876	195	XZ916	HAS2	16/2/79	30/3/79	w/off 13/10/88
WA877	196	XZ917	HAS2	1/3/79	5/4/79	w/off 6/3/81
WA878	197	XZ918	HAS2	3/4/79	6/6/79	to HAS 6
WA879	198	XZ919	HAS2	24/4/79	1/6/79	w/off 27/6/85
WA880	199	XZ920	HAS2	23/5/79	3/7/79	to SAR Mk5
WA881	200	XZ921	HAS2	17/7/79	20/9/79	to HAS 6
WA822	201	XZ922	HAS2	24/8/79	2/10/79	to HAS 6
WA883	—	IN551	Mk42A	23/11/79	10/3/80	to India
WA884	—	IN552	Mk42A	14/12/79	14/3/80	to India
WA885	—	IN553	Mk42A	11/2/80	28/3/80	to India
WA886	—	ZA105	HAR3	14/8/80	11/9/80	to RAF
WA887	—	ZA126	HAS5	26/8/80	2/10/80	to HAS 6
WA888	—	ZA127	HAS5	9/9/80	2/10/80	to HAS 6
WA889	—	ZA128	HAS5	13/10/80	26/11/80	to HAS 6
WA890	—	ZA129	HAS5	24/11/80	13/1/81	to HAS 6
WA891	—	ZA130	HAS5	7/1/81	3/2/81	to HAS 6
WA892	—	ZA131	HAS5	3/2/81	3/3/81	to HAS 6
WA893	—	ZA132	HAS5	19/2/81	7/4/81	w/off 12/5/82
WA894	—	ZA133	HAS5	3/4/81	14/5/81	to RN
WA895	—	ZA134	HAS5	28/4/81	2/6/81	to HAS 6
WA896	—	ZA135	HAS5	18/5/81	6/7/81	to HAS 6
WA897	—	ZA136	HAS5	17/6/81	5/8/81	to RN
WA898	—	ZA137	HAS5	17/7/81	3/9/81	to HAS 6
WA899	—	ZA166	HAS5	4/3/82	21/5/82	to HAS 6
WA900	—	ZA167	HAS5	22/4/82	28/5/82	to HAS 6
WA901	230	ZA168	HAS5	2/6/82	2/7/82	to RN
WA902	—	ZA169	HAS5	4/8/82	13/9/82	to HAS 6
WA903	232	ZA170	HAS5	2/9/82	21/10/82	to HAS 6
WA904	166	ZA290	HC4	26/9/79	10/12/79	ex WA870 w/off 18/5/82
WA905	167	ZA291	HC4	6/11/79	4/12/79	ex WA871
WA906	168	ZA292	HC4	20/12/79	11/1/80	ex WA872
WA907	169	ZA293	HC4	31/1/80	20/6/80	ex WA873
WA908	206	ZA294	HC4	10/4/80	29/4/80	w/off 19/5/82
WA909	207	ZA295	HC4	1/5/80	3/6/80	to RN
WA910	208	ZA296	HC4	28/5/80	20/6/80	to RN
WA911	209	ZA297	HC4	27/6/80	24/7/80	to RN
WA912	210	ZA298	HC4	21/8/81	1/10/81	ex demo G-BJNM

WESTLAND SEA KINGS — ALL MARKS — continued

C/N	Bld No.	Serial	Mark	F/F	D/D	Remarks
ZA913	—	ZA299	HC4	10/9/81	1/10/81	to RN
WA914	—	ZA310	HC4	16/9/81	1/10/81	to RN
WA915	233	ZA311	HC4	22/10/81	2/12/81	w/off 23/4/82
WA916	235	ZA312	HC4	18/12/81	26/1/82	to RN
WA917	237	ZA313	HC4	27/1/82	24/2/82	to RN
WA918	239	ZA314	HC4	20/9/82	2/10/82	to RN
WA919	228	QA30	Com3	14/6/82	26/11/82	to Qatar
WA920	229	QA31	Com3	27/7/82	14/1/83	to Qatar
WA921	240	QA32	Com3	8/4/83	20/5/83	to Qatar
WA922	241	QA33	Com3	13/5/83	21/6/83	to Qatar
WA923	242	QA34	Com3	8/6/83	22/7/83	to Qatar
WA924	243	QA35	Com3	15/7/83	8/9/83	to Qatar
WA925	244	QA36	Com3	23/8/83	22/9/83	to Qatar
WA926	245	QA37	Com3	5/10/83	4/1/84	to Qatar
WA927	236	ZB506	Mk4X	19/11/82	23/12/82	to R.A.E.
WA928	237	ZB507	Mk4X	19/1/83	4/2/83	to R.A.E.
WA929	238	N16-238	Mk50A	7/12/82	22/2/83	to RAN
WA930	239	N16-239	Mk50A	9/2/83	11/5/83	to RAN
WA931	246	ZD476	HC4	8/12/83	9/1/84	to RN
WA932	247	ZD477	HC4	19/1/84	2/2/84	to RN
WA933	248	ZD478	HC4	15/2/84	1/3/84	to RN
WA934	249	ZD479	HC4	7/3/84	29/3/84	to RN
WA935	250	ZD480	HC4	19/3/84	2/4/84	to RN
WA936	251	ZD625	HC4	18/4/84	1/5/84	to RN
WA937	252	ZD626	HC4	21/5/84	5/6/84	to RN
WA938	253	ZD627	HC4	25/6/84	5/7/84	to RN
WA939	254	ZD630	HAS5	9/8/84	13/9/84	to HAS 6
WA940	255	ZD631	HAS5	12/9/84	26/9/84	w/off 9/9/91
WA941	256	ZD632	HAS5	20/9/84	2/10/84	w/off 15/10/86
WA942	257	ZD633	HAS5	23/11/84	4/12/84	to HAS 6
WA943	258	ZD634	HAS5	21/1/85	1/2/85	to HAS 6
WA944	259	ZD635	HAS5	8/2/85	1/3/85	w/off 21/6/85
WA945	260	ZD636	HAS5	1/3/85	21/3/85	to RN
WA946	261	ZD637	HAS5	23/4/85	21/6/85	to HAS 6
WA947	—	ZE368	HAR3	22/5/85	3/6/85	to RAF
WA948	—	ZE369	HAR3	29/7/85	16/8/85	to RAF
WA949	—	ZE370	HAR3	21/8/85	5/9/85	to RAF
WA956	—	ZE418	HAS5	22/1/86	31/1/86	to RN
WA957	—	ZE419	HAS5	18/2/86	3/3/86	to HAS 6
WA958	—	ZE420	HAS5	9/4/86	29/4/86	to HAS 6
WA959	274	ZE421	HAS5	9/4/86	11/1/89	to RN
then later sold to Pakistan and converted to Mk45 (ZG-935/4516)						
WA961	—	ZE422	HAS5	10/7/86	24/7/86	to HAS 6
WA952	—	ZE425	HC4	10/9/85	24/9/85	to RN
WA953	—	ZE426	HC4	20/9/85	2/10/85	to RN
WA954	—	ZE427	HC4	14/11/85	4/12/85	to RN
WA955	—	ZE428	HC4	10/12/85	18/12/85	to RN
WA960	—	ZF115	HC4	3/6/86	3/7/86	to RN
WA962	—	ZF116	HC4	8/8/86	27/8/86	to RN
WA963	—	ZF117	HC4	19/8/86	8/9/86	to RN
WA964	—	ZF118	HC4	16/9/86	30/9/86	to RN
WA966	—	ZF119	HC4	13/10/86	30/10/86	to RN
WA967	—	ZF120	HC4	11/11/86	28/11/86	to RN
WA968	—	ZF121	HC4	2/12/86	18/12/86	to RN
WA969	—	ZF122	HC4	15/1/87	5/2/87	to RN
WA973	—	ZF123	HC4	25/3/87	3/4/87	to RN
WA974	—	ZF124	HC4	4/4/87	24/4/87	to RN
WA950	—	ZF526	Mk42B	17/5/85	10/12/90	IN532/IN513
WA951	—	ZF527	Mk42B	18/7/85		w/off 23/7/87 (IN514)
WA965	—	IN555	Mk42C	25/9/86	23/7/87	also G-17-31
WA970	—	IN556	Mk42C	8/1/87	23/7/87	also G-17-32
WA971	—	IN557	Mk42C	26/3/87	23/7/87	also G-17-33

WESTLAND SEA KINGS — ALL MARKS — continued

C/N	Bld No.	Serial	Mark	F/F	D/D	Remarks
WA972	278	IN515	Mk42B	7/5/87	17/1/90	also ZG601
WA978	292	IN514	Mk42B	28/7/87	13/4/89	also ZG602
WA979	294	IN516	Mk42B	7/10/87	20/3/90	also ZG603
WA975	—	IN558	Mk42C	26/8/87	1/11/88	also G-17-34
WA976	—	IN559	Mk42C	31/10/87	31/10/88	also G-17-35
WA977	—	IN560	Mk42C	5/1/88	1/11/88	also G-17-36
WA980	295	IN517	Mk42B	14/11/87	20/6/90	also ZG604
WA981	296	IN518	Mk42B	23/12/87	20/2/89	also ZG605
WA982	297	IN519	Mk42B	29/1/88	11/4/89	also ZG606
WA983	298	IN520	Mk42B	22/6/88	21/2/89	also ZG607
WA984	299	IN521	Mk42B	14/10/88	12/4/89	also ZG608
WA985	300	IN522	Mk42B	30/11/88	12/4/89	also ZG609
WA986	301	IN523	Mk42B	2/3/89	11/9/89	also ZG610
WA987	302	IN524	Mk42B	17/2/89	22/5/89	also ZG611
WA988	303	IN525	Mk42B	29/3/89	14/9/89	also ZG612
WA989	304	IN526	Mk42B	21/4/89	13/9/89	also ZG613
WA990	305	IN527	Mk42B	10/5/89	12/9/89	also ZG614
WA991	306	IN528	Mk42B	18/5/89	30/11/89	also ZG615
WA992	307	IN529	Mk42B	16/6/89	14/12/89	also ZG616
WA993	308	IN530	Mk42B	7/7/89	30/3/90	also ZG617
WA994	309	IN531	Mk42B	13/8/89	25/6/90	also ZG618
WA996	—	ZG829	HC4	10/4/89	3/5/89	w/off 20/10/92
WA1001	—	ZG820	HC4	2/6/90	21/7/90	to RN
WA1002	—	ZG821	HC4	7/8/90	24/8/90	to RN
WA1003	—	ZG822	HC4	18/9/90	15/10/90	to RN
WA997	—	ZG816	HAS6	7/12/89	9/1/90	to RN
WA998	—	ZG817	HAS6	1989	23/3/90	to RN
WA999	—	ZG818	HAS6	1989	9/4/90	to RN
WA1004	—	ZG819	HAS6	10/5/90	25/6/90	to RN
WA1005	—	ZG875	HAS6	27/6/90	24/8/90	to RN
WA995	—	IN533	Mk42B	11/4/90	11/12/90	also ZG980
WA1006	—	ZH566	Mk43B		Late 1991	to RNoAF

F/F = First Flight
D/D = Delivery Date
(CU) = RNAS Culdrose RNFTU

This list, although not complete, is a good guide to the production programme for all Marks of the Sea King and gives almost all the first flight dates. The majority of new build Royal Navy HAS Mk5s have now been uprated to the HAS Mk6 standard.

WESTLAND HELICOPTERS LTD
ROYAL NAVY SEA KINGS/RAF SEA KINGS

HAS Mk1 — 56 built: XV642 to XV677 (36 a/c), XV695 to XV714 (20 a/c/)

HAS Mk2 — 21 built: XZ570 to XZ582 (13 a/c), XZ915 to XZ922 (8 a/c)

RAF HAR 3 — 19 built: XZ585 to XZ599 (15 a/c), ZA 105 (1 a/c), ZE368 to ZE370 (3 a/c)

HC Mk4 — 41 built: ZA290 to ZA299 (10 a/c), ZA310 to ZA314 (5 a/c), ZD476 to ZD480 (5 a/c), ZD625 to ZD627 (3 a/c), ZE425 to ZE428 (4 a/c), ZF115 to ZF124 (10 a/c), ZG829 (1 a/c), ZG820 to ZG822 (3 a/c)

HC Mk4X — 2 built: ZB506/ZB507 (2 a/c) RAE Bedford/Farnborough

HAS Mk5 — 30 built: ZA126 to ZA137 (12 a/c), ZA166 to ZA170 (5 a/c), ZD630 to ZD637 (8 a/c), ZE418 to ZE422 (5 a/c)

HAS Mk6 — 5 built: ZG816 to ZG819 (4 a/c), ZG875 (1 a/c)

Total of 174 built for Britain

OVERSEAS WESTLAND SEA KINGS

Mk41 — 23 built for Germany: 89+50 to 89+71, 89+61 crashed prior to delivery and was replaced by a 2nd a/c.
Mk42 — 12 built for India: IN501 to IN512
Mk42A — 3 built for India: IN551 to IN553
Mk42B — 21 built for India: IN513 to IN533. IN514 crashed prior to delivery and was replaced by a second a/c. IN513 — one of the development a/c was brought up to production standard as IN532.
Mk42C — 6 built for India: IN555 to IN560
Mk43 — 10 built for Norway: 060, 062, 066, 068, 069, 070, 071, 072, 073, 074
Mk43A — 1 built for Norway: No.189
Mk43B — 1 built for Norway: No.ZH566
Mk45 — 6 built for Pakistan: 4510 to 4515
Mk45B — 1 sold to Pakistan: 4516 (ex RN HAS Mk5 ZE421/ZG935)
Mk47 — 6 built for Egypt: WA822 to WA827
Mk48 — 5 built for Belgium: RS-01 to RS-05
Mk50 — 10 built for Australia: N16-098, 100, 112, 113, 114, 117, 118, 119, 124, 125
Mk50A — 2 built for Australia: N16-238, 239
Mk1 Commando — 5 built as Mk70 for Egypt: 261 to 265
Mk2 Commando — 17 built for Egypt: 721, 722 WA806, WA808 to 821
MK2A Commando — 3 built for Qatar: QA20 to QA22
Mk2B Commando — 2 built for Egypt: VIP transport — 723, WA807
Mk42C Commando — 1 built for Qatar: VIP transportation (G17-26)
Mk2E Commando — 4 built for Egypt: ECM Helicopters — SU-BBJ, ARR, ART, ARP
Mk3 Commando — 8 built for Qatar: QA30 to QA37

150 Sea Kings built for export plus one Ex-RN HAS5

List compiled by Mr Bob Turner.

ROYAL NAVY

Sea Kings lost — these helicopters were total write-offs and not repaired and returned to service.

XV645	13/1/72	737NAS	Ditched off Portland — Loss of power
XV646	25/10/77	264H/814NAS	Ditched off HMS *Hermes* in the Western Approaches during exercises — Loss of yaw control
XV652	3/2/88	132/826NAS	Lost while operating from HMLMS *Poolster* in the Mediterranean
XV658	3/2/83	016N/820NAS	Ditched into the Atlantic off HMS *Invincible*
XV662	10/4/72	140E/826NAS	Ditched five miles off the Lizard
XV667	12/12/74	145TG/826NAS	Ditched off HMS *Tiger* and sank in the Bay of Biscay
XV668	24/2/87	586/706NAS	Ditched three miles off Dodman Point
XV695	17/11/75	303PW/819NAS	Lost while operating from HMS *Hermes* during exercise 'Ocean Safari' off Norway — Single engine failure
XV698	11/7/82	351/824NAS	'C' Flight — Double engine failure while on VERTREP to HMS *Leeds Castle* in South Atlantic
XV702	21/3/74	045R/824NAS	Flew into cliffs near Coverack on return from a night dunking exercise
XZ572	14/1/80	270/814NAS	Lost off the Scottish coast due to loss of engine oil pressure
XZ573	18/5/82	022/826NAS	Ditched into the South Atlantic due to instrument failure — later sunk by naval gunfire
XZ582	27/10/89	264N/814NAS	Ditched off Bermuda and sank — operating off HMS *Invincible*
XZ915	6/3/81	015N/820NAS	Collided with XZ917 off the Isle of Wight while operating from HMS *Invincible*
XZ916	13/10/88	130/826NAS	Ditched into the sea 15 miles off Plymouth during a night landing on RFA *Resource*
XZ917	6/3/81	see XZ915	
XZ919	27/6/85	129/826NAS	Collided with C-130 off the Falkland Islands
ZA132	12/5/82	134/826NAS	Lost five miles south of HMS *Hermes* in the South Atlantic — Single engine failure
ZA290	18/5/82	VC/846NAS	Burnt by the crew at Agua Fresca, 11 miles south of Punta Arenas, Chile
ZA294	19/5/82	VT/846NAS	Bird strike to the tail rotor South Atlantic
ZA311	23/4/82	VP/846NAS	Ditched during a night VERTREP mission on the way to the Falklands
ZD632	15/10/86	010R/820NAS	Ditched off Gibraltar while on approach to HMS *Ark Royal*
ZD635	21/6/85	703/819NAS	Crashed on Spears Hill near Tayport on the way from Dutch tanker *Leuchars*
XZ577	1/6/91	138/826NAS	Rotor blade strike to RFA *Fort Grange* during 'Operation Mana'. The aircraft sank and was not recovered
ZD631	9/9/91	266/814NAS	The first HAS Mk6 to be lost — Ditched 40 miles off Shetland while operating from HMS *Invincible*

SECTION 25

SIKORSKY S-61/SH-3

The S-61 (US Navy designation SH-3) gave the US Navy their first all weather hunter/killer ASW helicopter. Equipped with a Dunking sonar (AQS-10/13) and 840lb of weapon load (torpedoes or depth charges) the twin turboshaft engines and boat-shaped hull provided safety in over water operations.

The S-61 proved a success in both military and civil variants and licence production agreements were signed with Westland Helicopters Ltd in the UK, Agusta in Italy and Mitsubishi in Japan.

S-61/SH-3: First flew on 11 March 1959 powered by two General Electric 1,050shp T58-GE-6 turboshaft engines. Only a small number of SH-3s were built and the type entered service with the US Navy in 1961.

S-61/SH-3A: 1961–62 saw the introduction of the SH-3A with uprated 1,250shp General Electric T58-GE 8 turboshaft engines. These were built in a number of variants including the VH-3A/VIP, HH-3A/SAR and exported as the S-61A (transport) and S-61B (ASW). Within eight months the SH-3A captured all of the world's five major helicopter records. On 7 February 1962 it gained the helicopter speed record by flying at 210.6mph over a 19km course. Other US Navy variants on the SH-3A included the RH-3A used for mine countermeasures equipped anti-mine towing equipment. These were soon replaced by the more powerful CH-53.

S-61/SH-3D: These introduced more powerful 1,400shp General Electric T58-GE-10 turboshaft engines and uprated transmission increasing the gross weight to 20,500lb (9,299kg). This variant also incorporated improved mission equipment, also the VH-3D/VIP.

S-61/SH-3G/SH-3H: The SH-3G Utility and SH-3H ASW variant saw the Sea King's gross weight increase to 21,000lb. The ASW variant introduced mission equipment including uprated radar, variable depth sonar, active and passive sonobuoys, data link, chaff/flare, a tactical navigation system and magnetic anomaly (MAD) equipment.

There have been numerous variants of the S-61 used by the USAF, USMC and US Coast Guard in the transport, Search and Rescue (SAR) and Combat SAR roles. These include:

USAF: CH-3C/E and HH-3E (Rear cargo door and loading ramp).

HH-3E (Jolly Green Giant) combat SAR with Inflight refuelling probe. Two HH-3Es made aviation history in 1967 by flying non-stop across the Atlantic Ocean. They were refuelled in flight by four HC-130 tanker aircraft and set two new helicopter world records for the New York to London and New York to Paris routes.

US Coast Guard. In 1965 the US Coast Guard selected the HH-3F as their SAR helicopter. These were similar to the HH-3E with rear loading ramp and improved navigation and communications equipment.

The SH-3H and SH-3G are still operational with the US Navy and are used in the ASW, logistic support and Search and Rescue roles. The Sea King is now being replaced in the ASW role by the SH-60F Sea Hawk and this change should be complete by the late 1990s. The SH-3H will then be converted to the search and rescue utility roles for the US Navy.

CIVILIAN S-61

The S-61 proved highly successful in its civil variants and is today operated throughout

Below right: Sikorsky S-61N (G-BFMY) at RAF Mount Pleasant in the Falkland Islands. Bristow Helicopters have been operating two Sikorsky S-61Ns (G-BFMY/G-BCLD) in the Islands supporting the military since 1983. By 1990 Bristow S-61s had flown over 20,000 hours in the Falkland Islands. *(Patrick Allen)*

Below left: The first of nine Danish Air Force Sikorsky S-61A (U240) which are used in the SAR role. *(Patrick Allen)*

the world undertaking a variety of missions from logging to oil field support.

The S-61L was designed as a 25 to 30 seat passenger airliner and the S-61N for amphibious operations. The S-61L became the first twin turbine helicopter certified for passenger operations by the US Federal Aviation Agency in 1961 and began scheduled passenger service with Los Angeles Airways in early 1962. The S-61L was also the first helicopter to be certified by the FAA for Instrument Flight (IFR) and the S-61N for amphibious passenger operations. In 1969 Sikorsky introduced the S-61N Mk11 and S-61L Mark 11. These introduced more powerful 1,500shp General Electric CT58-1400-1/2 turboshaft engines, improved soundproofing, and new vibration absorbers. The Mark 11 was faster, more easily maintained, and was able to carry more passengers than the original Mk1s.

By 1970, the commercial S-61 was in regular scheduled passenger service with Ansett-ANA in Australia, British European Airways in the UK, Greenland Air in Greenland, Elivie in Italy and New York Airways, Los Angeles Airways and San Francisco and Oakland Helicopter Airlines in the USA.

The S-61N proved highly successful in the offshore oil rig support role and became the workhorse in the North Sea operating with British Airways Helicopters, Bristow Helicopter Ltd, Helikopter Service of Norway and KLM Noordzee Helikopters. Offshore oil drilling and production platform support saw the S-61 operating around the world. Brunei Shell and Bristow used S-61Ns off Malaysia, Helicopter Utilities flew S-61Ns to rigs off Northern Australia, Okanagan Helicopters' S-61Ns supplied rigs in Hudson Bay and off Canada's Atlantic coast.

BRISTOW S-61N Mk11 (SAR)

Today the S-61 still undertakes the bulk of offshore helicopter support around the world. Bristow Helicopters Ltd has recorded over five million passengers carried on half a million flights. Used primarily by the Company for North Sea offshore support operations a number of their S-61Ns are also configured for the Search and Rescue role. Operated in the dedicated search and rescue role (SAR) under UK Government contract for HM Coastguard, these aircraft are fitted with Bristow developed specialist SAR equipment including a fully-coupled Flight Path Control and 'auto-hover' system making them the most capable civil search and rescue helicopter in service in the world. Other equipment includes a nose-mounted Forward Looking Infrared (FLIR) and Bendix RDR 14000C colour, weather, mapping and search radar and 295ft variable speed rescue hoist.

FALKLAND ISLANDS

Bristow Helicopters Ltd have been operating the S-61N Mk11 in support of British Forces in the Falkand Islands since 1983. Originally operating three Sikorsky S-61Ns these were reduced to two in October 1983.

Of the original three S-61Ns, two aircraft G-BFMY and G-BCLD still remain and are in daily use by Bristow Helicopters Ltd based at RAF Mount Pleasant. The main task of the S-61Ns is to fly troops/passengers, mail, rations, water, fuel and other freight around the many outstations. Much of their tasking involves the external and internal carriage of stores, supplies and troops.

For operations in the Islands, the S-61Ns who usually carry 24 passengers are roled for

Below left: British Airways Helicopters S-61N (G-BFFJ) on a construction task fitting air conditioning units onto the roof of a supermarket. One of a variety of roles undertaken by the S-61N. *(Patrick Allen)*

Below right: US Navy SH-3H (148988) from HS-3 Squadron painted in the latest toned down colour. *(Patrick Allen)*

18 troops plus internal freight. External loads weighing up to 6,000lb can also be carried. By 1986 the Falkland-based S-61Ns achieved 10,000 flying hours and in 1990 this had exceeded 20,000 hours flying in support of British Forces. Contracted to 170 hours per month, to include air testing and training, the S-61Ns undertake VFR (Clear of cloud in sight of the surface) daylight only operations. The aircraft are fitted with basic navigational equipment which includes VOR, ILS and NDB. The Bendix weather radar can be used as an aid to navigation if required. In their support of the military, which occasionally includes deck landings, carrying dangerous air cargo, missiles and armed troops, Bristow Helicopters Ltd have obtained a CAA dispensation allowing them to undertake military operations. Minor and deep maintenance on the S-61Ns is undertaken in theatre, with maintenance working parties being flown in from the UK for major overhauls. Both aircraft have been fitted with a Helicopter Health and Usage Monitoring System (HUMS) and for their support of the military, VHF and UHF radios have also been installed.

SIKORSKY SH-3H SEA KING/US NAVY

Fuselage Length: 54ft 9in
Height: 73ft
Weight empty: 11,865lb
Max AUW: 21,000lb
Speed: 166kts max; 136kts cruise
Ceiling: 14,700ft
Range: 542nm
Powered by: Two General Electric T58-GE-10, 1,400shp turboshaft engines
Crew: Four (including two sonar operators)

SIKORSKY VH-3D SEA KING/USAF/USMC

The VH-3D is a VIP variant of the SH-3D equipped with air conditioning, custom interior and a special communication package. The first version of the VH-3A flew President Kennedy more than 25 years ago. These were replaced by 11 VH-3Ds with uprated engines, improved avionics and communication systems. These Sea Kings have recently been joined by 9 Sikorsky VH-60 Black Hawks operating with the Executive Flight Detachment who provide the transport for the President of the United States of America.

Fuselage length: 54ft 9in
Length: 72ft 8in
Height: 16ft 10in
Weight empty: 14,249lb
AUW: 20,500lb max take-off
Speed: 140kts max; 115kts cruise
Ceiling: 14,700ft
Range: 400nm
Engines: Two General Electric T58-GE-400B turboshafts
Crew: Three

Below: Italian Air Force Agusta HH-3F SAR helicopters are moving more towards the military Combat SAR role and the fleet is currently being updated. This includes a new low visibility grey colour scheme, NVG compatible cockpit lighting and the old Ecko 290M nose radar is being replaced by an improved APS-717 radar. The AFCS is also being updated along with an improved navigation and avionic suite which now includes Loran-C, RWR, Chaff/Flare plus Forward Looking Infrared (FLIR). *(Agusta)*

SIKORSKY S-61N MARK 11

Overall length: 73.00ft (22.3m)
Overall height: 19.00ft (5.8m)
Max AUW: 20,500lb (9,300kg)
Cruise speed: 110kts (204km/hr)
Max range: 365nm (676km) standard fuel tanks
Engines: Two General Electric CT58 140-2 1,500shp turboshafts
Crew/seating: Up to 24 passenger seats and two crew plus cabin attendant's seat

SIKORSKY BUILT S-61 SEA KINGS

Denmark: 9 × S-61As (U240, 275, 276, 277, 278, 279, 280, 281, 481)
Spain: 22 × SH-3D, SH-3G, SH-3H
HS 9-1 to 6 (D)
HS 9A-7 to 12 (G)
HS 9-13 (D)
HS 9A-14 to 17 (G)
HS 9-18 (D)
Note: A number of a/c have been converted to the AEW role with Type Designation HS.9L (HS.9L-11/HS.9L-12).
Israel: Unknown number of S61R/HH-3s delivered 12 known.
Argentina: (7 known) S-61D-4, H71, H72- Air Force.
0675 to 0678, 0696 (VIP fit) — Navy.
Malaysia: 40 × S-61A-4 NURI (By 1992 35 × S-61A-4 Nuris remained in service with the Royal Malaysian Air Force and are being upgraded by Airod Industries with new radar and avionics etc.
Indonesia: 1 × VIP transport S-61 known.
Brazil: 6 × S-61D-4/SH-3Ds N-3007 to 3012.
US Airforce: 133 × CH-3B, CH-3C, GCH3C, CH-3E and HH-3E.
US Navy/USMC: 247 × SH-3A, UH-3A, SH-3G, AH-3A, SH-3H, CH-3B, NSH-3A, HH-3A, SH-3D, VH-3D, YSH-3J.
US Coast Guard: 40 × HH-3F Pelican plus six ex Air Force CH-5Es.

Right: One of two VIP Agusta AS-61A. *(Agusta)*

Top right: An Italian Navy (MARINAVIA) Agusta ASH-3H ASW/ASUW Sea King armed with two Martel air-to-ship Sea Killer missiles in the ASUW role. Like the Royal Navy Sea King HAS Mk 5/6 these Sea Kings will be largely replaced by the EH101 Merlin in due course. *(Agusta)*

AGUSTA/SIKORSKY SEA KINGS

Peru: 9 × AS-61A-4s (ASH-3D/H).
Iran: 15 × AS-61A-4s (ASH-3D).
Libya: 4 × AS-61A-4s plus one VIP transport.
Saudi Arabia: 4 × AS-61A-4s.
Italy: 28 × ASH-3D- MM5003N to 5029N, MM5031N, MM5032N.
18 × HH-3F- MM80974 to 80993.
2 × AS61A- MM80972, MM80973.

Below: San Diego Bay, California. A diver of Helicopter Anti-submarine Squadron 10 (HS-10) jumps from the door of an SH-3H Sea King helicopter during search-and-rescue training. *(US Navy)*

UNITED AIRCRAFT OF CANADA SEA KINGS

Canada: 41 × CHSS-2 or CH124/SH3A — 12401 to 12441.

MITSUBISHI SEA KINGS

Japan: 123 × S-61B/S-61B-1 (HSS-2). 10 × S-61A-1s.

Below: US Navy Sea King SH-3G 149728 fitted with MAD unit and painted in the latest toned down markings. The SH-3H and SH-3Gs are due to continue in service for several years to come. The ASW SH-3H will be replaced by the SH-60F Sea Hawk. *(Sikorsky)*

Below: Pacific Ocean. A Helicopter Anti-submarine Squadron 8 (HS-8) SH-3H Sea King helicopter flies near the guided missile cruiser USS *Valley Forge* (CG-50) as the ship is underway with the aircraft carrier USS *Constellation* (CV-64) battle group. *(US Navy)*

SECTION 26

EH101 MERLIN THE NEXT GENERATION

During the Royal Navy Equipment Exhibition in September 1991 the go ahead was announced for a production order of 44 anti-submarine Merlin helicopters to be powered by Rolls-Royce Turbomeca RTM322 engines for the Royal Navy. It was also announced that the Westland/IBM team was to be responsible for the intregation of the mission systems with the airframe.

Merlin will give the Royal Navy a formidable all weather weapon system which will be capable of detecting and destroying silent, deep diving nuclear or conventional submarines at far greater distances than ever before. Capable of all weather, single pilot operations in VFR and IFR conditions, Merlin in the ASW role will have a cruising speed of 150kts and a five-hour endurance. The helicopter will be able to operate in the anti-ship (ASuV), airborne early warning (AEW), airborne surveillance, search and rescue (SAR) and the mine countermeasures roles. Configured for the SAR role Merlin can accommodate eight stretcher cases and 10 seated casualties.

Merlin makes extensive use of modern composite materials to help reduce weight and improve damage tolerances and fatigue

Below: First flight of EH101 PP1/ZF641 at Yeovil on 9 October 1987. *(Patrick Allen)*

to both fuselage and dynamics. This includes the rotor head and advanced aerodynamic rotor blades. Modern technology has also allowed fixed-wing standards of safety and incorporates triple hydraulic systems, crash-worthy fuel tanks, Active Vibration Control and a comprehensive Health and Usage Monitoring (HUMS) System for engines, transmission and structure. This allows real time health monitoring of critical components and helps reduce operating costs. The three RTM322 turboshaft engines develop 2,100 shaft horsepower and greatly improve power margins allowing Merlin to recover from the hover and transition away on two engines. The improved power flexibility of three engines also allows Merlin better cruise economy with one engine on standby. Initial delivery of Merlin to the MOD is due in 1996/1997.

Opposite: The first deck landing by the Royal Navy pre-production EH101 Merlin PP5/ZF649 aboard the Type 23 Frigate HMS *Norfolk* which took place on 15 November 1990. This was followed by a successful ship handling and interface trial in June/July 1991 and in September 1991 the Royal Navy confirmed a production order for 44 Merlins. *(Patrick Allen)*

EH101 PRE-PRODUCTION AIRCRAFT

EH101 PPI/ZF641. First flight 9 October 1987 (Yeovil)
EH101 PP2. First flight 26 November 1987 (Milan). (Crashed 28 January 1993).
EH101 PP3. First flight 30 September 1988 (Yeovil).
EH101 PP6. First flight 26 April 1989 (Milan).
EH101 PP4/ZF644. First flight 15 June 1989 (Yeovil).
EH101 PP5/ZF649. First flight 24 October 1989 (Yeovil/RN).
EH101 PP7. First flight 15 December 1989 (Milan).
EH101 PP8. First flight 24 April 1990 (Yeovil).

EH101 PP5/ZF649 is the dedicated Royal Navy pre-production aircraft and is fitted with a fully integrated mission and avionics system. This aircraft will also be used to develop the EH101's environmental control system and will fly the longer sorties needed to check long term running temperatures, communications systems and verify the aircraft's software. On 15 November 1990 PP5/ZF649 successfully undertook the first deck landings aboard the Type 23 Frigate HMS *Norfolk* and later undertook ship interface and handling trials which were also a complete success. This included testing deck lock and steering systems and Merlin's main rotor and tail fold system for storage in the ship's hangar.

Below: July 1992 and EH101 Merlin (PP5/ZF649) takes time out from its busy development programme to visit Yeovilton Air Day to show off its new main rotor 'Beanie' and Hub covers. *(Patrick Allen)*

ROYAL NAVY
ZF649
PP5
DANGER

EH101 MERLIN ROYAL NAVY ASW/ASuV MARITIME HELICOPTER

Dimensions

Main rotor: 18.6m (61ft)
Tail rotor: 4.0m (13ft 2in)
Length overall: 22.8m (74ft 10in)
Width: 4.5m (14ft 10in)
Height: 6.7m (21ft 10in)
Operating weight empty: 9,275kg (20,500lb)
Gross weight: 14,338kg (31,500lb)
Max speed Vne: 167kts
Range: 625nm
Time on station (ASW): 5 hours
Engines: Three Rolls-Royce/Turbomeca RTM322 turboshafts developing 2,100shp
Mission equipment: Ferranti Blue Kestrel 5000, 360 degree search radar; Racal Orange Reaper ESM system; GEC acoustic processor and communications Data Links; Passive and Active sonar systems with sonar buoys being launched from twin dispensers together with an advanced Active Dipping Sonar (ADS). Both systems are processed through the latest digital AQS903 processor. All of Merlin's mission systems and tactical data can be processed and displayed on Merlin's Mission Console.
Weapons: Up to four lightweight homing torpedoes or a mix of torpedoes and depth charges. Two missiles including Harpoon, Exocet, Sea Eagle and Martel Mk2 long-range anti-ship missile.

Below: EH101 Merlin flies over the Royal Navy Type 23 Frigate, HMS *Norfolk*. *(Westland)*

SECTION 27

'EXERCISE TEAMWORK 92' AND 'ORIENT 92'

Having deployed to the Gulf, Kurdistan and Bangladesh in 1991 the Sea Kings of both 845 and 846 Naval Air Squadrons based at RNAS Yeovilton spent their remaining months of 1991 being fully overhauled, repainted in their original 'junglie' green colour and brought up to the same operational standard to include TANS 2/GPS RWR, MAWS, Chaff/Flare IR jammers etc.

Having missed their winter deployment in 1990–91 both 845 and 846 Squadrons returned to RNoAF Bardufoss in December, January and February to undertake their Arctic Warfare and Flying training with the RN 'Clockwork' cell at Bardufoss. This period included refresher training and abinitio training. The NATO 'Exercise Teamwork 92' in March was the culmination of the winter training period for both 845 and 846 Squadrons with aircraft being deployed ashore and embarked aboard HMS *Fearless* and the RFA *Fort Grange*.

Prior to the winter deployment the decision was made to reduce the front line strength of both 845 and 846 Squadrons from eight to seven aircraft with each squadron sending a Sea King to 772 Naval Air Squadron at RNAS Portland.

'ORIENT 92'

On 12 May 1992 the Royal Navy's 'Orient 92' Task Force sailed from Portsmouth for a six and a half month deployment through the Mediterranean, to East Africa, the Far East and home via the Gulf. Lead by Rear Admiral John Brigstocke RN the Commander United Kingdom Task Group or COMUKTG with HMS *Invincible* as his Flag Ship the Group was made up of: HMS *Invincible*, HMS *Norfolk*, HMS *Boxer*, HMS *Newcastle*, RFA *Fort Austin* and RFA *Olwen*.

The deployment was to demonstrate the Royal Navy's ability to operate a Rapid Reaction Force and their continued ability to operate for prolonged periods outside the NATO theatre and to maintain and develop defence relations with the countries visited. The concept of the Royal Navy operating for prolonged periods away from home base was admirably proved in the South Atlantic in 1982, and during the Gulf War.

Below: 814 Squadron Sea King HAS Mk 6 (ZD633/167) cross deck landing aboard the RFA *Fort Grange* to collect priority stores during Exercise 'Teamwork 92' in northern Norway. An 846 Squadron Sea King HC Mk 4 is seen in the background waiting for the deck to clear. *(Patrick Allen)*

The Sea King helicopter continues to prove itself invaluable in increasing the flexibility and reach of this out of area Task Force with both ASW and AEW Sea Kings deployed aboard the Flag Ship HMS *Invincible* and the Commando Sea King undertaking the helicopter delivery service (HDS) and troop lift role aboard the RFA *Fort Austin*.

Left: Arctic camouflage painted 845 Squadron Sea King HC Mk 4 (ZF117/YD) lifting away from the RFA *Fort Grange* in Storfjorden Fjord. Like other front-line Commando Sea Kings (YD) has been fully enhanced with MAWS, RWR, GPS etc. Having deployed to Norway in early February 1992 to undertake the first 'Clockwork' training period since the Gulf War, the squadron deployed into the field during Exercise 'Teamwork 92' in support of the Royal Marines and Dutch Marines returning to the UK in late March. *(Patrick Allen)*

Below: 'Orient 92' and 814 Squadron Sea King HAS Mk 6s, 849 Squadron AEW Sea Kings and an 845 Squadron Commando Sea King are lined along the deck of HMS *Invincible* along with Sea Harrier FRS 1s from 800 Squadron. *(Patrick Allen)*

HMS *Invincible*
Sea Kings

7 × Sea King HAS Mk6: 814 Naval Air Squadron. (The Flying Tigers).

3 × Sea King AEW Mk2: 849 Naval Air Squadron ('A' Flight-Aardvarks).

RFA *Fort Austin*
Sea Kings

2 × Sea King HC Mk4: 845 Naval Air Squadron (YA-ZD477/YB-ZA312) ('C' Flight).

Right: An 845 Squadron Sea King HC Mk 4 along with a grey ASW Sea King HAS Mk 6 from 814 Squadron undertake a Vertrep (Vertical Replenishment) task between Gibraltar airport and RFA *Fort Austin* with a cloud covered Gibraltar Rock in the background. Over 40 loads of fresh food were flown aboard RFA *Fort Austin* who later that day rejoined the 'Orient 92' Task Force in the Mediterranean. *(Patrick Allen)*

Below: As the workhorse of the 'Orient 92' Task Group an 845 Squadron Sea King undertakes a Vertrep to the Fleet Support Tanker RFA *Brambleleaf*. *(Patrick Allen)*

SECTION 28

'OPERATION GRAPPLE' UN NAVAL SEA KING CASUALTY EVACUATION IN BOSNIA

On Wednesday 11 November 1992 four 845 Naval Air Squadron Sea King HC Mk4s embarked aboard the Royal Fleet Auxiliary Ship (RFA) *Argus* bound for the former Yugoslavia.

The Squadron's initial destination was the coastal town of Split in Croatia. Here they were to provide a dedicated United Kingdom national helicopter casualty evacuation capability in support of the 2,400 British troops, deployed to Bosnia-Herzegovina on 'Operation Grapple' supporting UN operations in the region.

The British troops, mainly from the Cheshire Regiment, form part of a 30 nations United Nations Protection Force (UNPROFOR) helping to protect and escort United Nations High Commissioner for Refugees (UNHCR) humanitarian aid convoys. These convoys were to distribute food and aid supplies to the thousands of refugees in the Bosnia and Herzegovina region already suffering the first winter snows.

In addition to the aircraft, some 90 men and 35 vehicles from both 845 Squadron and elements of CHOSC-Commando Helicopter Operations and Support Cell, also embarked aboard the RFA *Argus* along with four RAF Aeromedics joining the Sea King aircrew to undertake the Case-Vac/Air Ambulance role.

For their role in 'Operation Grapple', the Naval Sea Kings were painted in white UN livery and all were modified with full defensive measures consistent with the type of operation and combat environment they were to operate in. These enhancements were similar to those fitted for 'Operation Granby' and 'Operation Haven', the Kurdish relief operation and included:

Armoured crew seats; AAR-47 Missile Approach Warners (MAWS) now automatically linked to the M-130 Chaff/Flare dispensers; AN/ALQ-157 Infra red jammers; Updated Radar Warning Receiver (RWR); 'Have-Quick' frequency hopping UHF/VHF radio; Mode 4 IFF; GPS-TANS-2; 'Bright Star' Infra red floodlight and 'Grimes' dual infra-red/white light landing light. These two new items were fitted to help enhance Night Vision Goggle operations with aircrew issued with Nite-Op Gen 3 NVG goggles.

The Sea Kings were also equipped with 7.62mm cabin door guns for self protection. The aircrew plus RAF Aeromedics were issued with Kevlar body armour or Flak jackets and carried their SA80 rifles or 9mm Browning pistols. The aircraft were also equipped with specially adapted Arctic 'Pulks' usually carried in Norway, holding arctic/cold weather survival equipment should the helicopter go down in the mountains. All these enhancements increased the Sea King's all-up-weight (AUW) from around 13,700lb to approx 14,200lb. The normal max AUW for the Sea King HC Mk4 is 21,000lb increasing to 21,400 in cases of operational need. For their Case-Vac role, the Sea King can be adapted to carry six stretchers plus four seated casualties. In practice, particularly when operating in hostile territory, where aircraft are vulnerable to ground fire, stretchers and casualties are normally carried aboard and placed on the floor between the seats.

Below: Four 845 Squadron Sea Kings embarked aboard RFA *Argus* at Portland prior to sailing for Croatia on 11 November 1992. During the nine-day transit maintainers took the opportunity of fitting on the M130 Chaff/flares prior to flight testing all the new enhancements. *(Patrick Allen)*

Arriving in Split, Croatia on Thursday 19 November 845 Squadron immediately placed two of their aircraft at 45 minutes readiness to undertake 24-hour day/night Case-Vac missions in support of British troops undertaking UN relief operations in the Tomislavgrad, Vitez and Tuzla regions of Bosnia-Herzegovina, an area approximately 80 miles into the mountains to the east of Split. This area is extremely mountainous with high snow-covered peaks and the promise of winter snow at lower levels making ground transport along the narrow and twisting roads slow and hazardous. This was one of the reasons the Commando Sea Kings were deployed to the area to provide a quick response all-weather day/night Case-Vac capability able to take casualties out of the front-line and back to the nearest UN Field Hospitals. The Commando Sea King squadrons' capability to operate from a ship's deck plus their previous mountain and Arctic flying experience has yet again, proved invaluable.

Right: A pair of 845 Squadron Sea Kings return to the deck of RFA *Argus*. *(Patrick Allen)*

Initially operating from the deck of RFA *Argus*, anchored off the coast near Split airport, 845 Squadron moved ashore operating from the former Yugoslavian Army Divulje Barracks. From the day of their arrival off Croatia, the Sea Kings were operationally available ready to undertake day/night SAR/ Case-Vac missions in support of UK troops throughout the region. With no time limit set on the squadron's involvement in 'Operation Grapple' and the UN Protection Force committed to at least 12 months the squadron celebrated their 50th birthday on 1 February 1993 in Split.

'OPERATION GRAPPLE'

845 NAVAL AIR SQUADRON

4 x Sea King HC Mk4s
ZA314/YD, ZA313/YE, ZG820/YF,
ZA298/YG.

Right: A cockpit shot showing the newly fitted RWR and MAWS system plus Have Quick radio etc. *(Patrick Allen)*

Below: A Sea King launches a pair of flares with RFA *Argus* in the background. Depending on the threat the flare system can be programmed to launch two flares at a time or ripple them out hoping to lure away an IR guided missile. *(Patrick Allen)*